GROWING AND COOKING BERRIES

GROWING AND COOKING

BERRIES

By Mary W. Cornog

This book has been prepared by the Staff of Yankee, Inc.
Designed by Denzel Dyer
Illustrated by Ray Maher

Published 1980
Yankee, Inc., Dublin, New Hampshire 03444

First Edition

Copyright 1980, by Yankee, Inc.
Printed in the United States of America

Library of Congress Catalog Card No. 80-50184
ISBN 0911658-09-2

For Sarah

GROWING AND COOKING BERRIES

CONTENTS

INTRODUCTION

Anyone who has ever raised his own berries, then picked and eaten them fresh from his own patch knows the joys thereof. A few well-chosen, carefully tended plants can not only furnish heavenly eating fresh from the patch, but also provide ample stock for freezing, canning, jams, jellies, and preserves. Delving into these put-by stocks in midwinter is like opening a treasure of packaged summer.

Another big advantage of growing your own berries is economic. A glance at the prices placed by supermarkets on berries shipped in from other regions and nowhere near fresh will give added impetus to any dawning berry-growing impulses. Give in, give in. Few crops demand so little attention and return such profuse harvests as cultivated berries.

At the back of this book are listings of sources for Farmers' Bulletins, plants, and other pertinent information. Most of the nurseries and seed companies mentioned publish a catalog, and will send it upon request. The list is not exhaustive, but does purport to be a fair representation of nurseries across the country that send plants out by mail.

GROWING AND COOKING BERRIES

GROWING BERRIES

GENERAL DIRECTIONS

Gardening technology appears to be improving at the same rapid rate as all other technologies. More efficient means of cultivation, fertilization, and pest control are constantly appearing on the market. But, although these improvements in machinery and chemicals may make an enormous difference to large-scale producers, the home gardener growing berries for the average family of four will find them of little practical value. On the contrary, the best, most productive, most efficient, and easiest methods of growing berries continue to be the oldest and most technology-free.

Among the best fertilizers are still those that come from the compost heap and the manure pile. Mulch of leaves, straw, pine needles, or grass-clippings — whatever is decomposable and locally available in sufficient quantity — still outdoes plastic films. Regular pruning and maintenance of the patch still largely circumvent or defuse problems with diseases and pests that even heavy doses of chemical sprays do not always overcome (although in some cases of serious infection or infestation, chemicals judiciously used can provide the only means of saving the crop and, sometimes, the patch).

SELECTING THE PROPER VARIETIES

Numerous varieties of all cultivated species are currently accessible to home gardeners. Research into the development of new varieties proceeds apace; each year new strains emerge more resistant to diseases, more productive, longer-lived, tidier, earlier or later to ripen, more tolerant of drought or moisture or cold; smaller, larger, bushier, sparser, leafier, redder, blacker, purpler, yellower, greener. In short, whatever characteristic or combination of characteristics is sought is either already available or in the works. The precision — and the predictability — of plant breeding is marvelous and expanding all the time as the actual and potential genetic combinations increase with every new variety.

However, important differences still exist. One strain of strawberries, for example, offers resistance to verticillium wilt, another to red stele, a third to drought, a fourth to winter-kill. Such characteristics should be taken into account along with differences in flavor, color, time of ripening, suitability for eating, freezing, making into jam. And so on. The diversity of traits and combinations of traits seem endless, but choice is nevertheless possible. A few minutes of calm assessment will narrow the field, eliminating most varieties on quite practical grounds.

First, consider the area of the country where the crop is to be grown. Northern areas demand strains that are winter-hardy; southern areas, strains that are heat-tolerant and do not require complete dormancy. Moisture available to growing things is also an important factor. Is there a lot of rain, a moderate amount, or very little? Choose varieties of berry that do well under the specific condition. What about length of growing season? Northern growers would be foolish to choose a type that either blooms or ripens during times when frost normally threatens. Proceeding in this way, it is not difficult to select those varieties best suited to your garden. Final choices, then, can be made on the basis of other, less critical factors, such as taste, growing habit, or color.

Examination of several catalogs will help immeasurably in the selection process. Discussing the matter with experienced local growers or nurserymen will add invaluable insight. And when it comes time to buy, purchase plants from a reputable firm, whether local or mail-order, and be sure all plants are certified disease-free. In growing berries, it is crucial to start with plants that are guaranteed healthy.

A number of cultivated varieties for each type of berry is given at the end of each chapter.

The better and more regularly the plants are cared for, the healthier they will be, the bigger the yield, and the longer their lives. Below is a guide to the number of berry plants required by the home grower.

Berry	Yield/plant (quarts of berries)	Number of plants	Number of varieties required for successful pollination	Years from planting to production
Blackberries	4-8	8-10	1 or more	1
Blueberries	3-4	6-8	at least 2*	2
Elderberries	2-3	8	2*	2
Raspberries	1-1½	20-25 of each variety planted	1 or more	1
Strawberries	½-1/ foot of matted row	25 of each variety planted	1 or more	1

TABLE 1.
NUMBER OF PLANTS RECOMMENDED FOR FAMILY OF FOUR

*Not entirely self-sterile, but produces more and better fruit if cross-pollinated.

PLANTING SITE

Basic site requirements for all species included in this book are almost identical. Exceptions are the elderberry, which likes things wet, and the blueberry, which likes things sour.

In general, all do best in a place where there is ample moisture but good drainage. A gentle slope is ideal. To hasten blossoming and ripening, choose a slope that faces east or south. To retard development in order to avoid frost or conflict with other berries, choose a northern or western slope. Flat areas are perfectly acceptable as long as they don't collect standing water. Avoid plots that lie in a low spot or hollow where cold air settles; these attract frost like magnets. Untimely frost nipping blossoms or fruit can mean the loss of your whole crop.

Avoid planting near a stand of trees or wild berries. Trees, with their massive root systems, compete energetically with berry plants for nutrients and moisture. Wild berries can infect cultivated varieties with debilitating pests and diseases, so either plant cultivated types at least 500 feet from any wild varieties, or root out all wild specimens in the area. The latter, obviously, is by far the safer course. This caution does not apply in the case of elderberries.

Do not plant raspberries, blackberries, or strawberries in areas that have been used to grow tomatoes, eggplant, potatoes, peppers, or other strawberries and raspberries within the last three or four years. In the West, avoid land where these or okra, melons, mint, apricots, almonds, pecans, cherries, avocadoes, or roses have grown within ten years. All these plants are susceptible in varying degrees to soil-borne fungus diseases such as verticillium wilt and black root rot, and the fungus spores remain viable in the soil for years after the plants themselves have gone. For this reason, choose either virgin soil or soil that has been used to grow beans, peas, cole vegetables, or other non-threatening species.

If the area intended for berries is now in sod — that is, now covered with grass or weeds — best hold off on plans to plant berries for another year, time enough for the grubs and wireworms that are particularly fond of young berry roots to move out, and for the grass and weeds you will turn under to decay and enrich the soil. After tilling, plant the area with a crop such as rye grass, alfalfa, clover, or other legume. Plow this crop when grown back into the ground as a "green manure" that will provide valuable extra organic matter and nitrogen. You'll find more detailed information for each type of berry in the relevant chapter.

All berries do best in soil rich in organic matter. If the soil in your chosen site seems lacking in this respect, this can be easily remedied. Just till in generous amounts of rotted horse or cow manure, well-decayed compost, or peat moss, and your plants will thrive.

ORNAMENTAL PLANTING

If space is a problem, remember that a few types of berry plants can be grown as shrubs on your lawn. Bush blueberries are especially good for this. As you would any bushy shrub, plant them in decorative spots, remembering that blueberries can become quite large, and that they do need protection from birds. Yet even the netting that is the best deterrent to avian predators comes in discreet colors.

Both blueberries and brambles can be planted and groomed to form a hedge, as can the elderberry as well. Because an elderberry hedge will become exceedingly dense, plant the first few plants where their exuberant expansion will not interfere with other plants or activities. If space allows for only a small, well-behaved patch of elderberries, discipline the plants regularly with a power mower and pruning shears. And for all types of berries, growing them in pots, jars, barrels, or window boxes is an attractive option.

If, for some reason, the berry plants cannot be planted in their permanent location right away, two alternative means of storage are available. One is to water plants whose roots are encased in soil thoroughly, or wrap in sopping-wet paper towels those that are bare-rooted, and then enclose the entire plant in a plastic bag and store in the refrigerator. This method is best used if delay in planting is not more than a few days long.

If there will be a longer delay, the plants are best heeled in to the ground. To do this, dig a trench about 6 inches deep, stir in some peat moss or compost, wet down well, and gently set the plants in, tilting them at a 45° angle. Cover their roots with soil, water them, and then water regularly to insure their survival. Then, at planting time, simply remove the plants from the trench *carefully* and set in their permanent location.

SOIL pH

An important general principle of berry-growing involves pH testing and adjustment. The soil's pH value reflects its alkalinity or acidity, with a pH value of 7.0 representing neutral. Most berries do best in a slightly acid soil, ranging from 4.5 to 6.5, depending on the species. Blueberries, however, like their soil very acid. Except for the areas in the southwestern part of the country, acidic soils are the general rule in this country. But the range in acidity is extensive, and some berries and other plants are very picky. Testing the soil before planting and adjusting its pH as needed often avert later problems. Acidify soil that is too sweet (alkaline) by spading in sphagnum peat moss, rotted pine needles, rotted oak leaves, rotted sawdust, or cottonseed meal. You can also do this chemically by adding powdered sulfur, urea, ammonium sulfate, or aluminum sulfate — *carefully,* in limited quantities, following package directions. Sulfur is better for clay-to-ordinary soils; aluminum sulfate, for sandy soils.

Sweetening acidic soil is far more common and much easier than its opposite. Simply work powdered limestone into the soil at spring plowing or spread it on top in the fall or spring and let it leach in with the rain or snow. Don't use more lime than necessary, or the soil will end up too alkaline. Usually, 10 pounds of lime for each 100 square feet of garden raises the pH value a full point.

MULCHING

The use of mulches can improve most berry patches. Mulches conserve moisture in the soil, maintain an even temperature around the roots of the plants, contribute their nutrients to the plants' nourishment, and protect the plants from the rigors of frost and heat. Most mulches readily available tend to be slightly acid, and therefore must be counterbalanced by regular additions of lime. A year or two of regular soil testing will indicate how often and how much lime is needed. Either do your own tests with a kit that can be purchased at most garden supply stores, or take a sample of soil in to your local agricultural extension service and have them test it for you. The county agency will analyze the soil not only for pH value, but for type and fertility as well. This means that its basic nature — sandy, loamy, clayey, rocky, marshy, dusty — and its mineral content will be defined. Minerals tested for include nitrogen, phosphorus, potash, and trace elements. With this data, the home gardener can easily adjust values that are unsatisfactory, so that healthy plants will grow. Although not every soil can be turned into the ideal environment, even the heaviest, most barren plot can be adjusted so that berries will survive on it. Additions of organic matter rich in assorted minerals and/or chemical fertilizers, specifically designed to impart a specific component, accomplish this. Thorough analysis of the soil also indicates any need there may be for supplementary moisture or irrigation.

Whether you perform your own soil test, or have the county extension agency do it for you, your county agent will be glad to supply information regarding the specific additives and amounts of them needed by your garden. Follow his advice and thereafter test your soil regularly and maintain the desirable balance in your soil through regular application of the necessary types of fertilizer.

To delay blossoming and harvest time, leave the mulch on a little longer than necessary; spring growth, blossoming, and fruit formation will be commensurately retarded. If, on the other hand, earlier harvest is the goal, pull the mulch off during the day and replace it on those nights when frost threatens. Or substitute sheets, newspaper, or other more easily manipulated anti-frost coverings for the mulch when needed. Choose overcast days for the demulching time; that way there is no danger of sunburn damage to plants that have been in the dark all winter.

TABLE 2.
ORGANIC SOURCES OF COMPONENTS NECESSARY TO YOUR GARDEN*

Nitrogen	Phosphorus	Potash	Trace Elements
Well-rotted manure, especially horse and cow; use less of chicken or rabbit	Rock phosphate	Wood ashes	Usually present in sufficient quantity in other sources mentioned. If your
	Bone meal	Granite dust	county agent finds
	Manure (see under Nitrogen)	Greensand (ocean sand)	a drastic lack in your soil sample,
Bone meal			ask him for the
Bloodmeal			best remedy.

*All these substances are also available as commercial chemical preparations.

PRUNING BRAMBLE AND BUSH BERRY PLANTS

Regular pruning improves yield, cuts down disease, and keeps the patch, row, or bush tidy and manageable.

Pruning improves yield by culling out weak canes and suckers that divert energy from the stronger canes. Tip-pruning (cutting back the tips of new canes) further concentrates the plant's efforts on a sensible number of good-sized fruits, rather than on a multitude of smaller berries.

For brambles, after the season's berries have been picked and the fruit-bearing canes of that year are showing obvious signs of decline, prune these canes right off at the crown, haul them away and burn them. Leave blueberry canes intact; they bear fruit for many years. Prompt removal of spent canes markedly reduces the chances of viral or other infection of the new canes; good ventilation among a few healthy canes also helps prevent infestation by insect pests that delight in dark, moist surroundings.

PESTICIDES

Chemical sprays are commercially available to combat most pests and diseases that afflict berry plants. But there are also organic countermeasures for most problems, ranging from specially compounded sprays to ladybugs and praying mantises. However, there may well be times when nothing will work except a strong chemical preparation. Use good judgment and common sense in deciding which kind to apply, and use any additive as sparingly as possible. The object is simply to prevent or control a problem, not generate room for several new ones. If chemical sprays are used, follow directions carefully. And be absolutely sure to wash away chemical residue from the berries before eating, jamming, cooking, freezing, or using them in any way. This book refrains from specific recommendations for sprays, in part because the sprays themselves and restrictions on their use change frequently. Local extension services and/or local garden supply stores will have up-to-date information.

The Grower's Guide Tables in the sections to follow give directions specifically applicable to each type of berry plant covered in this book. Questions not directly answered in these tables will generally be resolved by referring again to this chapter.

THE BLACKBERRY FAMILY
(*Rubus variosa*)

The blackberry has historically flourished so abundantly in the wild that, rather than wondering how to cultivate it successfully in a garden, civilized man has far more often worried about keeping it out of "civilized" areas. References to the berry during the early days of this country describe the colonists' endless battles with the stubborn plant as the settlers strove to clear land for pasture and crops. Although they glisten with luscious black fruit in August, the trailing tendrils of the canes are at most other times of the year a thorny tangle draping every available fence row and roadside, encroaching wherever possible on crop or pasture lands. Many is the aimless rustic amble that has been curtailed or impeded by a blackberry thicket; the blackberry's notorious reputation as the thorniest of the brambles is well earned. (There are, however, on the market, *thornless* varieties of blackberry.)

The fruits of the brier go far to redeem it. It is hard to find a tastier berry for pies, jams, jellies, cordials, or wine. Add to this the sublimely heady fragrance of a blackberry thicket in bloom. There is simply nothing quite like it on a warm spring afternoon.

Picking the wild blackberry, too, is a memorable experience — the warm sun on back and shoulders, the sounds of bees taking their sup of the juicy fruit, the heavy lush smell of the moist earth on which the thicket thrives, the rich scent of dead-ripe and decaying fruit.

Blackberries are neither a new food nor one restricted to this country. Fossil remains and archaeological evidence witness that earlier peoples throughout the temperate regions of the northern hemisphere have long enjoyed its fruit. The first mention of the berry in literature appears about A.D. 1000 as a scholar's comment on a Latin phrase. Still later, Shakespeare uses the blackberry as the focus for an apt comparison: "If reasons were as plentie as Black-berries...." (I *Henry IV*, II, iv, 265).

Folk medicine has long touted the blackberry as a source of vitamin A. In addition, a tea brewed from the roots and leaves of the bramble served as an effective aid in controlling dysentery. And a bowl or two of blackberries eaten during the high-summer overindulgence in fresh corn, peaches, plums, and other laxative fruits and vegetables restores internal balance.

Although the wild blackberry's ubiquitous presence would seem to suggest that it can withstand any extremes of winter temperature, this is not at all the case. Canes taken from the wild and placed in unprotected but fertile locations generally succumb to frost damage. The wild brambles flourish in well-protected locations where their entire length is blanketed from the harsh winters by heavy snow cover. If domestication of wild canes is to be accomplished, this vulnerability must be taken into consideration.

Moreover, the wild berry transplanted responds unpredictably to soil and/or moisture conditions. A site ideal for cultivated varieties may generate only a poor or mediocre response from the wild plant, and it will often fail to fruit every year.

Transplanting wild stock is also not advised because a wild plant, itself largely immune, can infect less hardy cultivated brambles with virus or fungus diseases. Far wiser than attempting domestication of a wild plant is devoting your efforts to plants specifically bred for garden cultivation. Several different varieties are available, with a wide range of growing habits, flavor, hardiness, and even thorniness.

Three distinct types of blackberry (and its hybrid relatives) exist: erect or bush, trailing, and semi-erect. (The last is a cross between bush and trailing varieties, and has characteristics of both.) Hybrids of blackberry-raspberry stock such as the boysenberry, perfected by Mr. Boysen (wine red, trailing), the loganberry, by Mr. Logan (semi-erect plant, large fruits, good for wine as well as eating), the youngberry, achieved by Mr. Young, who crossed loganberries with dewberries (very productive, disease-resistant), and the nectarberry (a very juicy type, also called ollaliberry, and found in California) are not true blackberries, although they propagate by tip-rooting just as blackberries do.

HARVEST

Harvest blackberries when they are fairly oozing with juice. If fruits are picked before this soft stage is reached, they will be tart and tasteless. Once picked, store or process the crop quickly; freezing is an ideal way to preserve them quickly for later use. At any rate, don't let the fruit sit in the sun, or it will develop an unpleasant bitterness. Check the blackberry patch daily, and try to pick in the cool of the morning after the dew has evaporated.

TABLE 3.
GROWER'S GUIDE TO ERECT (BUSH) BLACKBERRIES

A. Selecting a Variety. Check fruiting time with early frost dates in your particular region. Select a variety that is hardy enough to winter over. Check that blossoming time coincides with optimal temperature conditions in your locale. Take note of seediness, thorniness, growing habit, productivity, and if possible taste the fruit of the variety or varieties on which you settle.

B. Site. Bush blackberries require a moist, well-drained site in full sun.

C. Soil. Best is a sandy loam, with pH between 5.5 and 7.5. Test your soil pH; add lime if required, then test again before planting.

D. Fertilizing. At planting time and each spring thereafter, fertilize your soil with well-rotted manure, compost, or 10-10-10 fertilizer. For best results, check with your county agent regarding your soil's requirements.

E. Planting. Time — as early as the ground can be worked in your prepared site. Set plants 4-5 feet apart in rows 6 feet apart. For each plant, dig a hole about a foot deep and the same in diameter, mixing well-rotted manure or compost with the soil removed. Spread roots of plant and drop them over a grapefruit-sized ball of fertilized earth in bottom of hole. Fill in and tamp down earth around plant, filling in the hole so as to hold the plant securely erect and eliminate all air pockets, covering the crown with an inch of soil. Water thoroughly and apply mulch.

F. Mulching. At planting time and early each spring thereafter, mulch with hay, wood chips, bark, leaves, or grass clippings (or a combination of these). Spread around the plant in a circle of 30 inches radius, 4-6 inches deep.

G. Pruning. At planting time, prune the canes of the plant you set to 6 inches above the ground. Each year thereafter, in early spring when buds are visible, cut out all excess canes or suckers, leaving 5-6 healthy, new, or primocanes per crown in addition to the 5 or 6 canes from the previous year that will bear fruit this season. When they are about 8 feet tall, cut back primocanes, or first-year canes, to 30-36 inches, or hip height, so that branching can occur. Prune lateral branches of fruiting, or second-year, canes so that no lateral branch is longer than 12 inches. In late summer or fall, prune primocanes back to 20-30 inches; cut spent fruiting canes off at crown or base.

H. Trellising. Not necessary with bush blackberries.

I. Cultivating. If mulch is *not* used, cultivate regularly to remove all weeds, but carefully. Do not scratch deeper than 2-3 inches into the soil or you will injure the shallow, but extensive root system. If mulch *is* used, simply pull any weeds hardy enough to raise their heads through the mulch.

J. Propagating. Two methods of propagating for erect bush varieties are suggested. One is to carefully uproot and transplant suckers. Once planted, treat as you did the parent plant during its first year; cut back to 6 inches after transplanting and proceed as outlined in this table for first-year plants.

Another method is to take a 2-inch root cutting in early spring from a 1-year-old or older healthy plant and bury it in 1-2 inches of rich soil. Treat the new plant resulting as described above.

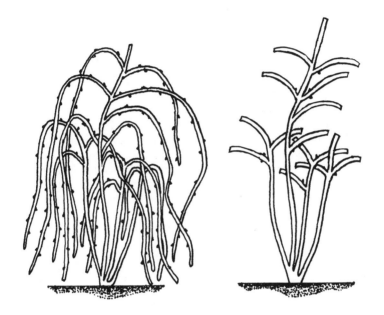

Pruning erect blackberries.
In spring, prune laterals that have grown on the previous year's canes back to 12 inches and let them blossom and fruit freely.

PROBLEMS

Insects and diseases to which members of the blackberry family are vulnerable correspond closely to those that affect raspberries. Assorted viruses, anthracnose, orange rust, and fungus producing double blossom can all infect the plants. The most effective precautions against disease are first, to buy only registered virus-free plants, and second, to root out and burn any infected plants, root and all, including wild brambles in the vicinity.

Insect pests that can potentially damage the plants include aphids, visible as greenish-brown clusters on leaves. The plants' leaves will curl and warp, eventually turn dry and brown. Aphids can be controlled by introduction of ladybugs into the patch or by applications of malathion.

Cutworms come in many species, ranging in size up to about 1¼ inches. Signs of their presence are the severing of healthy plants at or below soil level. They feed indiscriminately on leaves, buds, fruits, or roots. Control of cutworms is by affixing a cardboard collar around the stems of the plants, especially the freshly-planted, vulnerable tender young plants. One inch of the collar should be below the ground, two inches above.

Infestations of snails or slugs, often aggravated by wet weather, can be overcome by applying a little salt judiciously to their slimy bodies, or by setting out a shallow container of beer into which they will crawl and drown themselves. The skin of half a grapefuit or half a cantaloupe will also attract them, and they can then be scooped up en masse and disposed of.

Borers can also cause problems. The adult red-necked cane borer is a bronze-to-black beetle about an inch long. Its larva, the actual borer, measures about ¾ inch, is flat-headed, white, and slender. The beetles eat the leaves; the larvae tunnel into the pith of the canes. Evidence of the latter is the appearance of swellings along the surface of the canes. Control of this pest and other borers is best effected by cutting off infested canes below the swellings and burning them. But it cannot be reiterated too strongly that the best protection against virus, fungus, or insect infestation is a regular program of pruning and thinning, and prompt action if infection should occur.

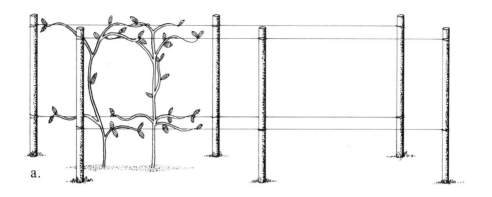

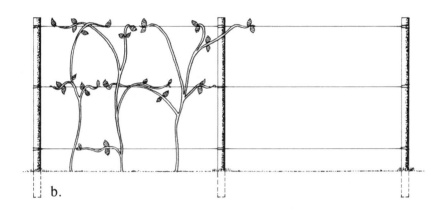

Trellising semi-erect or trailing blackberries.
(a). *Set posts at 8-foot intervals along both sides of row. Run two or three strands of wire between posts along rows to form a supporting trellis.*
(b). *Set a single set of posts along each row, run two or three strands along the posts, and train the canes around the wires.*
(c). *Set a post next to the crown of each plant and tie second-year fruiting canes loosely to it at several points.*

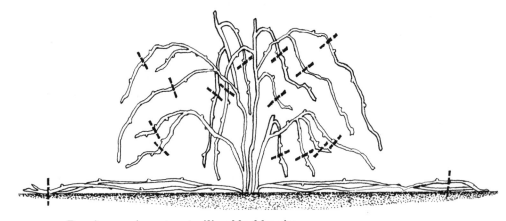

Pruning semi-erect or trailing blackberries.
Prune primocanes back to four feet and let trail along the ground. Prune laterals on second-year canes back to 12 inches for maximum yield.

TABLE 4.
GROWER'S GUIDE TO TRAILING OR SEMI-ERECT BLACKBERRIES AND HYBRIDS
Follow the directions outlined in Table 3, with the following exceptions:

A. Selecting a Variety. Not as hardy as bush or erect blackberries, few trailing or semi-erect varieties are suitable for northern climes, so check climate zones carefully if you live where winters are hard.

B. Planting. New plants should be set 6 feet apart in rows 8 feet apart. Prune new plants to 6 inches above the crown as for erect or bush types.

C. Pruning. Early next spring, when buds are visible, prune new canes so that each plant has only 12-16 canes. Tip back these primocanes when they are about 8 feet long to hip height to encourage branching. Prune the laterals of second-year fruiting canes to 12 inches long, and gently tie the canes to whatever type of trellis you choose (see below). Let primocanes trail along the ground until the following spring, when you will prune them as outlined above for second-year canes. Once the berries have been harvested from fruiting canes, prune these canes back to the crown; remove and destroy the old canes.

D. Trellising. Because of the tip-rooting habit of these types of berry, trellising of some kind is necessary to prevent rampant propagation. Trellis these the spring of the second, or fruiting, year. Either sink a stake close to each plant and tie fruiting canes loosely to it, or drive stakes at each end of the row and at intervals along it. Run wires at 2 or 3 levels along these stakes (along either one side or both sides of the row) and tie the canes loosely to them. (Let primocanes lie along the ground.)

E. Mulching. As in Table 3, but also: in northern areas, where frost damage is likely, cover the recumbent canes with mulch in the fall to protect them from the cold. Remove this covering in spring when all danger of frost is past.

F. Propagating. Trailing or semi-erect dewberries, loganberries, boysenberries, young-berries, and nectarberries are tip-rooters. The trailing tips of the canes put out roots if left to lie on the ground, or if covered with soil. Once roots are established, simply sever the cane 6 inches from the new roots and transplant. Treat as a first-year plant.

VARIETIES

Perusal of nursery catalogs will give the most comprehensive idea of the varieties available and best suited to a given locale. What follows here is simply a general guideline, not necessarily a set of specific recommendations.

Erect

Comanche, Cherokee, Early Harvest, Ranger, Raven, Tree Blackberry, Lawton — not hardy enough to withstand prolonged or severe winters.

Darrow, Ebony King, Eldorado, Lowden, Snyder — hardy in northern regions. Lowden is also orange rust-resistant.

Semi-erect

Black Satin, Dirkson, Evergreen, Smoothstem, Thornfree — none hardy in northern winters, all thornless.

Trailing

Boysen, Austin, Carolina, Brainerd, Cascade, Lucretia, Marion, Ola-lie, Youngberry, Logan — none hardy enough to grow in North.

BLUEBERRIES
(Vaccinium corymbosum)

Few other North American plants equal the blueberry in range and popularity. Whortle, hurtle, whin, blae, trackle have all been used to identify this one plant and its fruit, whose close relatives, wild or domestic, include bilberry, tangle- or dangleberry, and the renowned huckleberry. All produce fruit — the distinctive blue globe inextricably associated with the blueberry. The dusty blush on the fruits' skin is in reality a special strain of yeast that flourishes there. The huckleberry, however, is not a true blueberry; instead of the numerous tiny seeds found in a blueberry, the huckleberry has large hard seeds. It is a smaller plant than the blueberry, and its fruit lacks the blueberry's sweetness. Both species are members of the heath family, Ericaceae, to which azaleas, rhododendrons, and mountain laurels also belong.

A native of the North American continent, the blueberry grows wild anywhere in conditions acceptable to it, and such conditions exist in well over two thirds of the country: Maine west to Michigan, Wisconsin, and Minnesota; thence south to Florida, Mississippi, and Louisiana; and along the West Coast from Washington to Northern California. The critical factor is the acidity of the soil. Although they tolerate a wide range of acidity — pHs from about 4.0 to 5.5 — blueberries do best at a pH level of 4.3 to 5.0. Thus they will "make do" in many areas, but flourish abundantly in others. Only in the alkaline areas of the West and Southwest are they altogether absent.

The Indians relied in olden days on the annual blueberry harvest for a staple food they ate either fresh or dried. Dried, it was an important part of their winter diet. The dried berries are still used as a delicious substitute for dried currants or raisins.

Cultivated blueberries are distinctly a twentieth-century development, initially the result of efforts by F. V. Coville and Elizabeth White in New Jersey. Cultivated varieties now abound, and blueberries are today a major crop in many areas (e.g. Maine, New Jersey, Michigan).

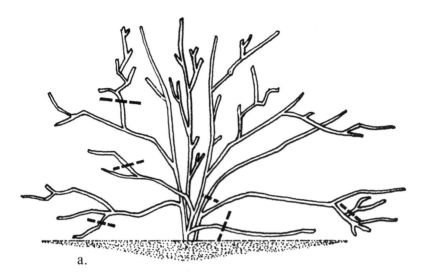

a.

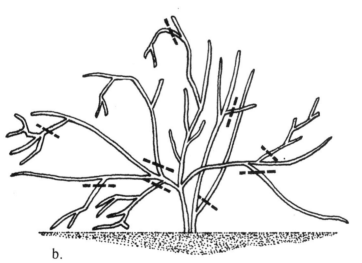

b.

Pruning blueberries.
(a). *During the plant's first six years, each spring save the three or four strongest new canes. Cut off all weaker new canes, canes growing too close to the ground, and all spriggy or twiggy growth on older canes.*
(b). *After the sixth year, keep the five or six strongest young canes and cut out the three or four oldest canes. Also remove excess new canes, including those growing close to the ground, and all spriggy or twiggy growth.*

Unlike blackberries, blueberries produce very differently under cultivation than they do in the wild. Cultivated plants are bigger and stronger, their berries bigger and more numerous. Although some contend that the flavor of cultivated varieties is bland compared to the wild type, in fact flavor varies widely from one variety to another. (Even in the wild, flavor will vary almost as noticeably from one bush to another.)

Cultivated berries conform to three types: highbush, lowbush, and rabbiteye. The first type grows throughout the middle and western range of the blueberry's habitat; the second, almost exclusively in Maine (on a semi-wild basis); and the third, from North Carolina southward. Cultivation and soil requirements are roughly identical for all three.

Although grown primarily as a fruit-producer, the blueberry makes a marvelous ornamental for hedges and general landscaping. Its attractive color — lush green in spring and summer, fiery red in fall — and lush growth make it ideally suited for the purpose.

In areas where pH levels are above 6.5, roughly the maximum level that can be counteracted, blueberries *can* be grown if conditions are closely controlled. Each plant must be set in its own soil environment, the contents of which are carefully monitored. A tub, such as the bottom half of an old drum, a washtub, or barrel, with holes drilled in the bottom to provide drainage, makes an excellent home. Sink it into the ground and fill it with a mixture of three parts sand, three parts sphagnum peat moss, and two parts acid leaf mold such as composted oak leaves or other high-acid matter, adjusted to the proper pH level. A blueberry planted in this and given annual feedings of acid fertilizer, whether organic or chemical, should thrive.

Care must be taken that chlorosis does not develop. Chlorosis has occurred if the leaves turn pale green or yellowish. It is usually the result of too much alkalinity, probably caused by sweetness leaching in from the alkaline soil surrounding the pot or the sweet water used to keep the plants moist. Counteract it as soon as it is noticed with applications of iron chelate, vinegar, or sulfur to leaves and soil. Healthy dark green color should return to the leaves in short order.

HARVEST

Harvest blueberries 7 to 10 days after berries have turned dusty blue, and when they are easily removed from the stems. Test berry clusters for ripeness by gently rolling the berries between your fingers. Those that come loose are ready to pick. Pick at 4 to 5 day intervals.

PROBLEMS

Potential hazards to blueberry plants include "mummy berry," a fungus disease that causes shriveled, dry berries; stunt virus; stem canker; botrytis or gray mold blight; powdery mildew; or witches-broom virus. There is a fungicide specific for mummy berry, but little beside proper pruning and care can be done about the rest. Canes or plants showing signs of fungus or virus infection should be cut off at the crown or rooted out altogether, removed from the patch, and destroyed.

Foremost among insect pests is the blueberry fruitfly. Its eggs are laid in the berry, and hatch to produce maggots and rot; control the problem with applications of rotenone from June through the end of harvest. Cultivation of the earth surrounding the plants will destroy fruitfly larvae pupating in the ground. But cultivate carefully to avoid damaging the plants' shallow roots.

Cultivation is also an effective control for the plum curculio, a small beetle that also lays its eggs in the green berries, leaving as its calling card a small depression. The cranberry fruitworm, a green caterpillar about one-half inch long, spins a web around a cluster of berries. Disrupt its pupating stage by cultivation; pick the worms off affected bushes by hand and destroy them.

Two effective biological controls exist for the cherry fruitworm, more often a problem to the commercial rather than the casual grower: the parasitic fungus *Beauvaria bassiana,* which attacks the hibernating larvae; and the *Trichogramma minutum* wasp, which preys on the adults.

Scales, whether Putnam, terrapin, or other, can be prevented by proper pruning and/or the application of a dormant oil spray in early spring. Problems caused by blueberry bud mites in the South can also be fended off by use of an oil spray in late September or early October.

The leaf miner, one of very few pests that chooses anything but the berry as its target, can most easily be controlled by plucking off affected leaves and

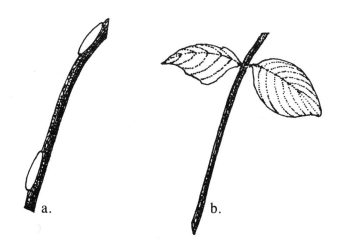

Blueberry cuttings for propagation.
(a). *Hardwood cuttings, taken from older canes, should be about 4 inches long, with a flat leaf bud just below the top cut and just above the bottom cut.*
(b). *Softwood cuttings, taken from new lateral growth, should be about 5 inches long, with a pair of leaves at the tip.*

TABLE 5.

GROWER'S GUIDE TO BLUEBERRIES
Highbush and Rabbiteye

A. Selecting a Variety. Select at least two that are suited to your planting area and climatic region. Plant varieties with different fruiting times for constant supply.

B. Site. Prepared patch, or lawn area (when used as ornamental shrub or hedge).

C. Soil. Acid, pH between 4.0 and 5.5, with 4.3-5.0 optimal. Sandy loam enriched with well-rotted compost or sphagnum peat moss.

D. Planting. Early spring, while plants are still dormant. (In mild climates, may be late fall.) Plants with earth around their roots preferable to plants with bare roots. Soak bare roots in a bucket of water for an hour before planting. Set plants 4-8 feet apart (set hedge plants only 3 feet apart) in rows 8-10 feet apart in holes 2 inches deep with 12-inch diameters. Mix a little acid (well-rotted oak leaves, pine needles, or sawdust), compost, or sphagnum peat moss into earth from each hole. Plant to soil level slightly above nursery mark (the joint between roots and crown that marks a previous soil level). Pack soil around plants, water thoroughly, and pack again. Mulch. Water weekly for 6-8 weeks. New plants 1-2 feet tall produce in about 3 years.

E. Mulching. Four to six inches deep at planting time. Additional can be added in late fall, but pull it back a few inches from the stem. Blueberries need cold to induce dormancy. In early spring, draw mulch up around stem again.

F. Pruning.
New plants: Prune back newly set plants over 12 inches high by ⅓-½. Pluck all blossoms the first spring. This procedure encourages a few strong canes with many fruiting buds. Blossoms are plucked in order to develop vigorous plant before serious fruiting. Early the second spring, cut out spindly branches and laterals and any dead wood.
Older plants: Starting the sixth spring of your plant's life, and each spring thereafter, prune out: a) all dead or broken branches and any clusters of thin, spriggy growth; b) branches close to the ground; c) 5-6 year-old canes, leaving only 3-4 strongest (mature

---

destroying them. However, it is rarely present in threatening quantities and does not compromise berry production.

If chemical control seems justified, check with the local agricultural extension agency to find out current guidelines.

Chlorosis, discussed on page 30 in conjunction with pH conditions, may be symptomatic of a number of other conditions, of which pH is the most common and most easily remedied. If pH inadequacy is not the cause, further investigation is warranted. Lack or superfluity of moisture or other nutritional deficiencies are the most likely culprits. Treat accordingly.

A common hazard to the harvest are your feathered friends, who will do their darnedest to gorge themselves on the entire crop of berries. The only way to keep birds away is to erect a barrier of cheesecloth, wire, or the netting designed specifically for this purpose. In a small patch, simply drape the net over the plants. The open mesh keeps birds out but lets sunlight, moisture, and air in. In a patch that contains more than just a few plants, it is more convenient to erect a framework of posts and wire around the plants on which

canes, "hardwood," have greyish bark; younger canes have greenish bark); d) new canes, leaving only 5-6 of the strongest. After pruning, plant should have 8-10 strong, well-budded canes with several strong laterals each.

G. Fruiting. Second spring, remove all blossoms to strengthen plant before serious fruiting. Third spring, allow plant to fruit. Yield modest at first, increasing each year until plant matures, about 8 years old. Fruiting may then continue for 30-40 years with good care.

H. Fertilizing. Yearly, with commercial acid fertilizer (5-10-5 or 7-7-7, designed for azaleas and rhododendrons) or with about 2 cups dried blood, bone, fish, or soybean meal, dried sheep manure, rock phosphate, cottonseed meal, or castor pomace — in early spring and later, about the time berries begin to form. Sprinkle around plant starting 10 inches from crown in a 3-4 foot radius. Granite dust or rock phosphate every 4-5 years.

I. Propagating.
Hardwood cutting (about 50% successful): With a sharp knife, take a 4-inch clean-edged cutting from mature growth in early spring so bottom cut is below a leaf bud, top cut just above a leaf bud. (Leaf buds are thin, flattish swellings along the branch. Flower/fruit buds are fatter.) Set cutting in rooting medium or one-half sand, one-half sphagnum peat moss so that top leaf bud is about 1 inch above surface. Water and enclose pot and cutting in clear plastic bag or wrap. Open plastic occasionally to water and let in air. Keep cutting in this protected environment until following spring.

Softwood cutting (little less than 50% successful): When first berries turn blue, take a 5-inch piece of new lateral growth. Strip all but 2 leaves at tip and proceed as with hardwood cutting.

Cane-rooting (easy, but chancy): In early spring, bend down low cane to ground and mound sand or soil/peat moss mixture over a stretch of cane to hold it down. Leave tip uncovered. Keep mounded covering moist. Roots will sprout from covered stretch of cane. Next spring, sever the cane between mound and mother plant, creating new plant.

Division (some risk, but the best bet): As in dividing any plant, dig up and divide in two, and replant. Water carefully to counteract shock to root system.

netting can be hung. Remember to provide a means of access to the crop. Hang the netting early in the season to discourage feathered predation from the outset.

LOWBUSH BLUEBERRIES

Although many of these varieties have much in common with the highbush and rabbiteye, they do demonstrate some idiosyncrasies not shared by the other two. Most marked is their essential wildness. Growing low to the ground — usually topping out at 6 to 24 inches — lowbush blueberries carpet scrub or barren areas of their own volition. Once having voluntarily seeded themselves in such an area, plants spread rapidly.

VARIETIES
Highbush *(Vaccinium corymbosum)*
Early: Bluehaven, Bluejay, Bluetta, Collins, Earliblue, Ivanhoe, June, Northland, Spartan, Walcott, Weymouth

Mid-season: Atlantic, Bluecrop, Blueray, Berkeley, Collins, Concord, Elizabeth, Meader, Patriot, Pemberton, Rancocas, Stanley

Late: Burlington, Coville, Darrow, Dixi, Elliott, Herbert, Indigo, Jersey, Lateblue, Rubel

Highbush especially suited to Maryland, Virginia, North Carolina — semi-hardy and/or canker-resistant
Early: Angola, Harrison, Morrow
Mid-season: Croatan, Scammell, Walcott
Late: Murphy

Rabbiteye *(Vaccinium ashei)*
Early: Aliceblue, Beckyblue, Climax, Flordablue, Gardenblue, Harrison (highbush), Premier, Sharpblue, Tifblue, Woodard
Mid-season: Bluebell, Bluegem, Briteblue, Delite, Homebell, Powder Blue, Southland
Late: Avonblue, Centurion, Menditoo

Lowbush *(Vaccinium augustigolium, myrtilloides)*
Low Sweet, Black-Fruited Low Sweet, Sourtop

Ornamentals, bred primarily for foliage and growing habits, but also good bearers; excellent for hedges and landscaping:
Evergreen Blueberry, Novemberglow, Ornablue, Top Hat

ELDERBERRIES
(Sambucus nigra)

Judging from scattered references in ancient tomes, elderberries were familiar to Greeks, Romans, and other early Europeans. Then as now one of the primary uses was in the making of wine. Judas Iscariot is said to have hanged himself on the branch of an elder. People in certain parts of Germany and Britain believe that Christ's cross was made of elder wood and insist that it is proper and necessary to doff the hat before an elderberry thicket.

The edible elderberry, *Sambucus nigra,* is native to Europe, North Africa, Western Asia, the Caucasus, southern Siberia, and eastern North America. (An inedible elderberry with a blue berry grows in the western United States.) It was and is highly prized as the source of medicines. The oil contained in the flowers, as an essence, flavors or scents candy, perfume, or lotion. A green extract of leaves imbues fat or oil with a pleasant green color.

Practical homeopathic uses are many, and involve all parts of the plant above the ground. Tea brewed from the bark serves variously as a cathartic, an emetic, a diuretic, and as a cure for infants' colic or for headache. Juice of the berries, taken fresh, is used for toothache in some parts of the world, for coughs in others. A salve is made from flowers and bark to heal insect and tick bites, ease pain of gout, burns, or tumors, and reduce swellings of assorted kinds. Leaves, with their distinctive odor, repel bugs, and keep a house free from ants and flies. They have been used as a compress to draw out infections. Berries of the plant were said to cure dropsy, rheumatism, and other swellings. Carried in the pocket, they were said to prevent poison ivy, and hung around the neck, to cure epilepsy. Various manipulations with a stick of elder wood were believed to soothe teething pain, eradicate warts, exorcise fever, cure rheumatism, deafness, sore throat, insanity, sleeplessness, depression, and hypochondria. But beware of using elder wood for a punitive switch: a recalcitrant child beaten with it will have his growth stunted.

In places other than Denmark, where superstition forbids it, the wood of

TABLE 6.

GROWER'S GUIDE TO ELDERBERRIES

A. Selecting a Variety. Largely self-sterile, so either plant at least two cultivated varieties near each other, or plant a cultivated variety within range of a wild thicket. Blossoming times of the two varieties should approximate each other.

B. Site. Moist, but well-drained in full or almost full sun. Plant to east of a building to gain protection from hard winters.

C. Soil. Loamy, with plenty of organic material.

D. Planting. In early spring, mix in well-rotted manure or compost with soil from planting hole. Dig holes 18 inches deep and 18 inches in diameter. Set plants 5 feet apart in rows 8 feet apart.

E. Mulching. Six inches deep, apply at planting time and every fall and spring thereafter (straw, compost, leaves, grass clippings, et al.).

F. Fertilizing. Almost never needed, aside from that provided at planting time and through mulchings.

G. Pruning. Growth habit very vigorous — can be grown as hedge. Clip elderberry suckers and seedlings when small with lawn mower by mowing every 2 weeks around your plants to keep them within bounds.

Plant grows to be about 6-10 feet tall, somewhere between a shrub and a tree in behavior. First-year growth is green and pliant; second- and third-year growth is woody, stiff, and fruit-bearing. After bearing the third year, the branch will normally die. Early next spring, prune away all such dead branches. Crown produces new shoots every year. Prune all but 5-6 strongest. You can tip these back to about 12 inches to encourage branching, but yield is so prolific that this is usually not necessary.

Generally, prune a bearing patch to keep it clean, uncluttered, and airy; prune a hedging patch minimally to keep it thinned but protective. Always prune in early spring while plants are dormant.

H. Propagating.
Division: Dig existing plant up, roots and all, divide, and replant separately.
Suckers: Transplant suckers put forth in early spring.
Self-seeded plants: Can also be dug up and transplanted.

the elderberry has long been valued as material for combs, skewers, mathematical tabulators, lathework, and pegs, especially the pegs used by shoemakers. Although the wood of young trees is brittle and pithy, more mature wood becomes attractively hard, pale, and close-grained, producing exquisite objects when worked and polished, and is well-suited for small work.

In addition to providing an essence used to flavor and scent other substances, the blossoms of the elderberry are a gourmet's delight, dipped in sweet batter and fried. Soaked in a jar of sugared water, they render a delicate thirst-quenching drink, called "elder blow," which can either be consumed as it is or fermented and drunk as a sweet, light wine. The wine produced from the ripe berries, and justly famed for its flavor, has long served as a delicacy in its own right or as a delicate additive to port wine.

HARVEST

Elderberries bear frothy white clusters of small blooms in late spring. These stand upright, are about 6 inches across, and appear flat across the top.

Berries appear shortly after blossomfall; they are first green, then red, turn purple, and finally black. The berries grow in clusters, like the blossoms. But this time the clusters are pendant, weighed down by the juicy heft of the berries. Pick elderberries when fully ripe, black but still firm, in mid- to late August. Remove the entire cluster, then later strip the berries from their stems with a fork or gentle fingers. Be sure to remove as many stem pieces as possible; leaving them in when cooking or otherwise using the berries distorts the flavor disagreeably.

PROBLEMS

Insects and diseases that afflict the elderberry are extremely few. A virus or two may crop up, but the incidence is rare. If disease should strike, of course, quickly root out and destroy infected plants, and watch others carefully for signs of sickness. Insects rarely, if ever, attack a patch. There seems to be a good deal of truth to the folk-belief in the elderberry's insect-repelling ability. The main problem with elderberries is birds, who love them. These plunderers can be deterred to some extent by strips of aluminum foil hung in the patch to flutter in the breeze, or aluminum pie plates hung there for the same purpose. Netting draped over the patch will keep the birds out, but since the plants frequently grow as tall as ten feet, applying a net shield is not always easy. However, a healthy elderberry thicket will bear plenty for birds and humans, and often begins prolific bearing the first year it is planted.

VARIETIES

At present, cultivated varieties available for home gardeners include:

Adams #2 — a type that produces berries earlier than most, available for planting in areas with short growing seasons.

Ezy Off — a good bearer with berries easier than most to strip from stems.

Johns — a bearer of larger but fewer berries than Adams, and requiring a longer season.

Kent — early season berry from Nova Scotia.

New York 21 — a mid-season bearer of many large berries, one of the best varieties available.

Nova — bred from a wild Nova Scotian strain, producer of extremely heavy and quite early crops.

York — late-ripening producer of bigger-than-average berries on large plants.

All varieties are available as certified virus-free plants. (Despite the elderberry's general immunity to viruses and other diseases, buying indexed virus-free plants is always advisable.)

RASPBERRIES
(Rubus idaea)

Many consider home-grown raspberries the absolute best of all. Soft and perishable as raspberries are, ripe berries freshly picked from the backyard bed are light-years better than those you can buy. To be at the peak of flavor, raspberries must be picked and eaten when fully ripe. Supermarkets, because of their requirements for shipping, storage, and shelf life, must stock fruits picked green and left to ripen en route. The full flavor never develops under refrigeration, and mold and mildew all too often do. Therefore, truly tasty raspberries can only be found in one's own private patch, or in a generous neighbor's.

Throughout history, the raspberry has symbolized the remorseful aspects of passions influenced by Venus. Some South Pacific cultures also associate the raspberry with death, and hang fronds of the plant over the doorway of a house where someone has died. The brambles, they believe, will entangle the dead person's spirit, should it try to reenter the house.

A close relative of the rose, raspberries of one sort or another are native to the temperate zones of Europe, Asia, and America. A wild Korean variety from temperate Asia is an important ancestor of many modern-day hybrids. Raspberries were known to the Greeks and Romans: the scientific name of the plant, still used today, *Rubus idaea,* means the raspberry that grows on Mount Ida, the site in Crete where, supposedly, Zeus was born. The berry did not come to be associated with Zeus, however, but with Venus and the sorrow that follows misplaced love.

Midwives in countries where the raspberry grows have traditionally used a tea brewed from raspberry leaves to relieve the pain of childbirth. During World War II, experimenters discovered a "new" drug called fragerine that eased the pain of childbirth without causing addictive or narcotic side effects. Made from a liquid brewed from dried raspberry leaves, fragerine is still used today. Raspberry vinegar, made by steeping a quantity of raspberries in vinegar, straining the liquid, and sweetening it slightly, relieves symptoms of flu and sore throats, acting as both an astringent and a stimulant.

Sometime in the early nineteenth century, botanists became interested in breeding better strains of raspberry. Early experiments were random and independent, such as the crossing done by Fannie Heath in Minnesota, which produced a new strain of black raspberry with bigger, juicier fruit and fewer thorns.

In the wild, and of course under cultivation, too, raspberries come in black, purple, red, and yellow varieties, and all manner of shades and crosses in between. Purple raspberries are a cross between black and red; the distinctively sweet yellows are a genetic mutation of red. Black and red, therefore, are the true genotypes at the root of all other strains. Red/yellow varieties differ from black/purple ones in manner of growth and propagation / and require different planting and pruning procedures (see Grower's Guide Tables 7 and 8). Red raspberries, furthermore, come in standard and "everbearing" varieties. The everbearers are called that because, although second-year canes bear crops the second season as usual, the primocanes bear late their first season.

Raspberries of all colors differ from blackberries and their relatives both genetically and in the relationship of berry to core. In the raspberry, the core (the torus or receptacle) remains on the plant when the raspberry is picked. In blackberries, the core comes away from the plant with the berry. In other respects, however, such as shape of leaf, composition of berries — round or elongated, composed of several grains or drupes — growing habits, or thorniness, the two species are hard to tell apart.

Principles of raspberry cultivation closely match those described for blackberries, as comparison of the Grower's Guide Tables will show.

Raspberries tend to do well in cold climates, provided adequately hardy varieties are selected at the outset. The raspberry blossoms slightly later than many berries, so that late frosts that can wreak havoc on other berries usually fail to nab the late-blooming raspberry.

The new plants come bare-rooted or set in a small ball of earth. In the former case, soak the plants immediately in a bucket of water, and let them sit in the water for a couple of hours. Then, plant as quickly as possible in the permanent location. As planting proceeds, keep waiting plants wrapped in wet burlap or other saturated material to prevent them from drying out. If the roots come encased in earth, keep the roots and soil moist while preparing to plant. Then set roots and soil into the earth "as is."

Standard red and yellow raspberries send up suckers prolifically, giving the grower a wide choice in selecting strong new canes to replace old ones that have borne fruit and died. Because of this habit, red and yellow plants tend to form a solid hedge, rather than a series of discrete clumps, and it is best to design their patch with this in mind.

In the case of standard varieties, a season's primocanes will form fruiting buds (plumper than the flatter, more numerous leaf buds) in the fall, hold them dormant over the winter, and produce blossoms and berries from them the next spring. "Heading back," or tip-pruning, the primocanes forces them

TABLE 7.
GROWER'S GUIDE TO RED/YELLOW RASPBERRIES

A. Selecting a Variety. Check fruiting times with early frost dates in your particular region. Select a variety that is hardy enough to winter over. Check that blossoming time coincides with optimal conditions in your locale. Consider both standard and everbearing varieties. Buy only stock certified as virus-free, since all types of raspberries are prone to infection. If possible, taste the fruit of the varieties under consideration.

B. Site. Moist, well-drained and slightly elevated, with some shelter and full sun.

C. Soil. Best is a sandy loam, with pH between 5.5 and 6.5. Test your soil pH; add lime if required, then test again before planting.

D. Fertilizing. At planting time and each spring thereafter, fertilize your soil with well-rotted manure, compost, or 10-10-10 fertilizer. For best results, check with your county agent regarding your soil's requirements.

E. Planting. Time — as early as the ground can be worked in your prepared site. Set plants 3-4 feet apart in rows 6-8 feet apart. For each plant, dig a hole about a foot deep and the same in diameter, mixing well-rotted manure or compost with the soil removed. Set plants with earth encasing their roots directly in the hole — as is. Spread bare roots and droop them over a grapefruit-sized ball of fertilized earth in bottom of hole. Fill in and tamp down earth around plant, filling in the hole so as to hold the plant securely erect and eliminate all air pockets. The crown should be lightly covered with soil. Set bare-rooted plants 2-3 inches deeper than nursery mark. Pack again. Water every 2-3 days for several weeks until plants are firmly established.

F. Cultivating. Cultivate frequently to control weeds for the first half of the plant's first season. Work the soil to a depth of 2-3 inches only, in order not to injure the shallow root system. About mid-July or early August, mulch (see below). Continue to remove all weeds hardy enough to poke through the mulch.

G. Mulching. In mid-July or early August after planting, and early each spring thereafter, spread good organic mulch around the plant 4-6 inches deep in a circle of 30 inches radius. Use hay, wood chips, bark, leaves, or grass clippings (or a combination of these). Replenish mulch in late fall.

H. Pruning.
Standard varieties: At planting time, prune the canes of the plant you set to 6-8 inches above the ground. Each year thereafter, in early spring when buds are visible, cut out all excess new canes or suckers, leaving 3-4 primocanes per crown and one sturdy sucker every 6 inches. When primocanes are about 4 feet tall, cut them back to 30-36 inches, or hip height, so that branching can occur. Prune lateral branches of fruiting canes back to 4-6 fat fruiting buds, or about 12 inches. In late summer or fall, prune primocanes back to 3-4 feet; cut spent fruiting canes off at base. Remove all prunings from patch and burn them.
Everbearing varieties (can be handled in two ways):
1. For two small crops, prune laterals of second-year canes in spring. Thin primocanes to 4-6 per plant and let grow. Second-year canes will produce fruit in midsummer. Primocanes will produce in late summer or early fall. In late fall, cut off and destroy spent second-year canes. Cut primocanes back to 3-4 feet in late fall, then treat as second-year canes the following spring.
2. For one large crop in late summer, mow off the entire patch at about 2 inches in early spring. Primocanes will soon sprout. Thin to 6-8 per crown and one sucker each 6 inches. Let grow. They will bear abundant fruit in late summer or early fall.

I. Propagating. These varieties send up suckers prolifically. Carefully uproot and transplant suckers. Once planted, treat as you did the parent plant during its first year; cut back to 6 inches after transplanting and proceed as outlined in Table 3, J, for first-year plants.

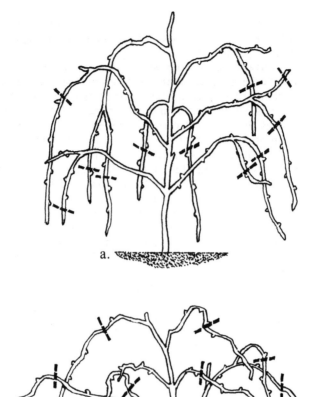

Pruning raspberries.
(a). *For a bountiful crop of standard red raspberries, prune laterals of second-year canes back to 12 inches in early spring.*
(b). *Prune black raspberry canes their second year, cutting laterals back to about 12 inches, leaving four or five fat fruiting buds on each.*

to develop laterals, which effectively doubles the number of fruiting buds per cane. Cutting back the laterals on the second-year canes encourages the production of fewer and bigger berries.

Primocanes of everbearing varieties will produce fruit their first year. To reap two crops from everbearers, select the six to eight strongest new canes in early spring, as for standard varieties. Then, rather than heading them back as you do standard canes, simply let them grow. Prune laterals of second-year canes in early spring as for standard varieties. These will blossom in spring and fruit in midsummer. Primocanes will blossom in July and produce ripe berries in August or September. After they have borne fruit, cut the primocanes back to 3 feet so they will branch. Cut off and discard spent second-year canes, but save primocanes. They will fruit again the following summer. The following spring, repeat the selection and pruning process.

It is also common practice to manage everbearers to produce only one large crop. To do this, thin the primocanes each spring and let them fruit. Then, early the following spring, mow them all off at a height of 2 inches. A new stand of primocanes will soon develop. Thin and proceed as before. This way, the bed produces its late crop only, and only on primocanes. The single late crop will be larger than either of the crops when both sets of canes are allowed to fruit.

If you plan to grow both standard and everbearing varieties, keep them segregated in distinct areas of the patch. This makes managing two pruning and fruiting schedules much easier.

With all raspberries, whether red/yellow or black/purple, proper pruning and thinning to allow free circulation of air among the plants significantly reduce the chance of fungus disease.

Over the years, black raspberries have been bred to grow in almost all regions of the country. Before their genetic makeup was taken under scientific management, however, the range was fairly restricted, covering only the mountain and piedmont areas of Pennsylvania and West Virginia.

In planting black and purple raspberries, it is wise to separate them from any red or wild varieties by a distance of 500 to 1000 feet. Diseases that hardly affect reds, yellows, or wild varieties can wipe out the more sensitive black and purple varieties.

HARVEST

Pick all varieties of raspberry at the peak of ripeness, when they are lush and juicy. Ripe berries are very easily bruised, so use shallow containers for the picking. Pick in the cool of the morning, after dew has dried, or in the late afternoon/evening before it has formed. Eat, process, or freeze as soon after picking as possible to capture maximum flavor. Pick daily or every other day; raspberries ripen and go by quickly.

PROBLEMS

Diseases and insects affecting raspberries are similar to those attacking other brambles and berries. Viruses include mosaic, leaf curl, orange rust, and streak, all of which produce symptoms that give the viruses their names and which often resemble nutritional deficiencies. Diagnosis, therefore, is difficult, but the slightest suspicion of infection is reason enough to dig up and destroy the entire plant immediately. Viral infections are systemic, and cannot be cured simply by excising the part that displays the symptom — the entire plant, including roots and suckers, must be eradicated. Keep aphids, the primary carriers of these diseases, out of your patch by inviting in some ladybugs; they love to eat aphids.

Anthracnose, also known as "gray-bark," hits black raspberries the hardest. This damaging blight appears on new canes as reddish- to dark-purplish spots that gradually expand and turn gray, from the center outward. Afflicted canes eventually become crusted with a gray film and are extremely prone to winterkill. Anthracnose is most effectively warded off by planting certified disease-free plants; by pruning and removing spent canes properly; by maintaining good ventilation through excision of weeds and excess plant growth; and by choosing varieties bred for resistance. These include: in black raspberries, Blackhawk and Dundee; in purple, Sodus; and in red raspberries, Cuthbert, Indian Summer, Latham, Newburgh. An additional preventive measure is application of a lime-sulfur spray in early spring, although organic gardeners frown on this resort to chemicals. Captan spray, applied according to directions, will also give some protection.

Crown gall causes bumpy growths on the roots, dwarfed and deformed canes, seedy fruit, and, ultimately, the death of the plant. There is no cure, and very little prevention: buy disease-free plants. If crown gall should appear, root out and destroy affected plants, and do not plant that particular area with raspberries, blackberries, or strawberries for three years.

Spur blight is the result of a fungus infection, but is not usually very serious. Symptoms include the sudden appearance of purplish blotches along the new canes, generally near the burgeoning laterals. Like most fungus infections, this one is encouraged by excessive moisture and/or poor air circulation.

Like crown gall, verticillium wilt, an organism that lurks in the soil, gains entry to raspberry plantings through the roots. There is no cure. Because it also afflicts tomatoes, potatoes, eggplant, and peppers, do not plant raspberries where these vegetables have been grown. Buy verticillium-resistant varieties.

The numerous insects that appreciate the succulence of raspberries can be successfully combated if infestations are caught early enough.

Borers come in three varieties: cane, root or crown, and red-necked. The first produces sawdust on the cane, and can thus be easily spotted, tracked down, and cut out. The root or crown borer causes the infested cane to wilt, and produces gall-like growths on the crown. Dig out the borer, and cut off and burn the wilted cane. The red-necked borer produces longish swellings and/or galls near the base of the cane; under the swelling lies the borer. Slit the cane, dig out the borer, or cut and burn the cane. If borers are present, it is doubly important that any nearby wild brambles or wild roses be demolished, since they, too, harbor the bugs.

TABLE 8.
GROWER'S GUIDE TO BLACK/PURPLE RASPBERRIES
Follow the directions outlined in Table 7, with the following exceptions:

A. Selecting a Variety. Not as disease-resistant as red, yellow, and wild varieties, so separate from those by 500-1000 feet.

B. Planting. Plant in early spring. In frost-free areas, can be done in late fall. New plants should be set 5 feet apart in rows 6 feet apart. Set in the ground as for red and yellow types, placing plants so soil level is 1-2 inches higher than in nursery. Water, and prune new plants back to 6 inches.

C. Pruning. Pinch off tips of new canes of first-year plants when they are 18 inches high so they will branch. Early next spring, when buds are visible, thin new canes so that each plant has only 4-5. Remove prunings. Prune tips of primocanes when they reach 18 inches. Prune back the laterals of second-year fruiting canes to 5 or 6 fat fruiting buds. Once the berries have been harvested from fruiting canes, prune these canes back to the crown; remove and destroy the old canes.

D. Trellising. Some black and purple varieties have canes sturdy enough to hold their heads off the ground without support. But others require trellising of some kind to prevent rampant propagation from tip-rooting. Trellis these the spring of the second, or fruiting, year. Either sink a stake close to each plant and tie fruiting canes loosely to it, or drive stakes at each end of the row and at intervals along it. Run wires at 2 or 3 levels along these stakes (along either one side or both sides of the row) and tie the canes loosely to them.

E. Mulching. As in Table 7, but also: in northern areas, where frost damage is likely, lay the canes down, bending them gently at the crown, and cover with mulch in the fall to protect them from the cold. Remove this covering in spring when all danger of frost is past.

F. Propagating. These plants are tip-rooters. The trailing tips of the canes put out roots if left to lie on the ground, or if covered with soil. Once roots are established, simply sever the cane 6 inches from the new roots and transplant. Treat as first-year plant.

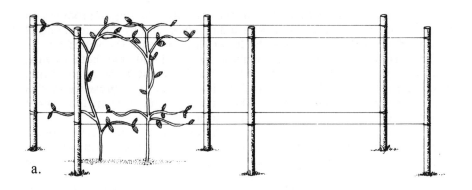

a.

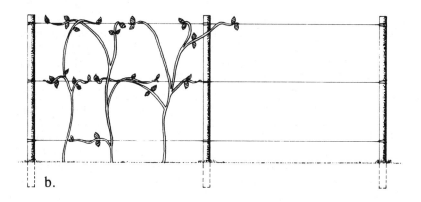

b.

c.

Trellising black or purple raspberries.
(a). *Run two or three strands of wire along pairs of posts set at 8-foot intervals along the row to support canes.*
(b). *Twine canes around two or three strands of wire along a single set of posts set 8 feet apart along each row.*
(c). *Or, while primocanes trail along the ground, tie second-year canes to a post set near the crown of each plant.*

The cane maggot can be spotted by the characteristic rings it produces on the cane, one above and one below its hiding place, and by the wilted end of the cane. Cut off the cane below the lower ring and burn the wilted tip.

The raspberry fruitworm presents a double threat, but both threats, adult and larval, can be counteracted by regular applications of rotenone, following package directions. The adult fruitworm shows its presence by slits in the fruit buds and new leaves. The larvae that hatch from the eggs it deposits then feed on the berries as they form.

The rose chafer, which also dines on roses, enjoys the leaves, buds, flowers, and fruit of the raspberry. It is a small tannish beetle with a slightly red head and long legs. It is usually not present in large numbers, and can easily be removed by hand. Squish it.

Infestations of red spiders, which may be either green or red, can be detected by the actual presence of spiders on the bottom sides of leaves, or by the presence of yellow specks and/or small webs on the leaves. The eventual result of a plague of red spiders is retarded plant growth and deformed fruit. Their hold is tenuous, however, and they and the dust that encourages them can be washed off with a good spray of water. They can also be combated with malathion or an all-purpose spray containing sulfur.

In many parts of the country, Japanese beetles are a terrible problem. They particularly delight in luscious young leaves, leaving behind only the veins. A serious infestation can gobble up the healthiest planting. All-purpose orchard spray, dormant oil spray, or milky spore disease, a natural control, are the best cures for their ravaging presence. The county agricultural agent can give up-to-date advice on combating them. Some varieties of raspberry seem less attractive to the beetle than others; for example, Success purple raspberry suffers little damage.

White grubs, the larvae of the noisy june bug, are most often a problem in soil that has recently been in sod, another reason for letting a year elapse before planting.

VARIETIES
Varieties of raspberries available for home cultivation become more numerous by the year.

Red Standard
Anelma — a new variety that ripens early to mid-season, produces many suckers, bears large berries, and is very hardy.

Boyne — a new variety rapidly becoming a particular favorite.

Canby — a thornless variety not noted for flavor, but easy to pick.

Chief — producer of early, small berries, recommended for the growing conditions in the Midwest.

Cuthbert — an old variety still popular, with tasty berries but light crop.

Earlired — one of the earliest to ripen.

Fairview — recommended for the less exacting winters of the Northwest.

Gatineau — good for the Northwest.

Haida — also designed for the moderate Northwest.

Latham — the most popular variety; seems to thrive everywhere.

Liberty — a new hardy variety, highly touted, from Iowa State University.

Newburgh — a hardy variety that produces firm berries on upright canes, so is popular with commercial growers; resists mosaic virus better than Latham.

New Hampshire Red — a hardy variety bred to ripen well before the end of the North's short growing season.

New Hilton — improvement on earlier Hilton; produces very large berries in mid-season; bushy erect plants; respectably hardy.

Ottowa Red — bred for demanding Canadian growing conditions.

Polaris — new hardy variety; good producer.

Reveille — very early and very hardy variety.

Southland — designed to bear well in the hot summers of the South; doesn't need freezing temperatures to induce dormancy.

Sumner — bred for the gentle conditions of the Northwest.

Sunrise — an old variety with delicious flavor; ripens mid-season; vigorous, hardy, productive, erect.

Viking — productive and hardy.

Willamette — great for the Northwest.

Red Everbearing

August Red — earliest of the double-croppers; must be mowed clean each spring to produce; fills gap between standard and fall-bearing varieties. Not a true everbearer, it bears its crop on primocanes only.

Durham — bred for the North; excellent flavor, very hardy, does not bear well if mulched.

Fall Red — disease resistant; bears large summer and fall crops; bears fall crop earlier than most.

Heritage — most popular everbearer; good crop.

Indian Summer — bred especially for the South and Northwest.

Prestige — a new variety; thornless, sturdy, vigorous; excellent crop; hardy.

Scepter — reliable and tasty.

September — late to ripen but good crop.

Yellow

Amber — hardy and very sweet.

Fall Gold — sister to Fall Red; everbearer; resistant to frost because of high sugar content of berries.

Golden Queen — unique in that it tip-roots like black raspberries.

Goldenwest — developed for and grown in the Northwest.

Black

Allen — the offspring of Cumberland and Bristol, gives large fruit.

Blackhawk — hardy and disease resistant.

Bountiful Giant — a productive everbearer.

Bristol — bears in mid-season; good quality crop.
Cumberland — reliable and popular.
Dundee — ripens late.
Huron — ripens late; resistant to anthracnose; does well in poor soil.
Jewel — new; disease-resistant.
John Robertson — very hardy, with large fruits.
Lowden — hardy and disease-resistant.
Morrison — ripens mid-season.
Munger — ripens mid-season.

Purple

Amethyst — disease resistant and hardy.
Brandywine — new; late; fairly hardy.
Clyde — probably the best; hardy and disease-resistant.
Sodus — tall, hardy; produces fruit over a long period.
Success — early; disease-resistant and unattractive to Japanese beetles.

STRAWBERRIES
(Fragaria)

The strawberry — plant, blossom, and fruit — has always connoted purity and passion. The symbol, therefore, has quite logically been used in conjunction with stories and pictures of the life of Jesus, dating from the early medieval period onward. Virgil and Pliny the Elder, Latin authors of the pre-Christian era in Rome, mentioned the berry in their poetic and historical works. In *Othello,* Shakespeare decorated the pure Desdemona's handkerchief with a symbolic strawberry.

A Norse tale related how Frigga, wife of Odin, played on Odin's fondness for the berries to smuggle the souls of dead infants into heaven. Odin forbade those souls entry, but Frigga concealed them in ripe strawberries and thus sneaked them into the divine precinct.

In parts of Bavaria, countryfolk still practice the annual rite each spring of tying small baskets of wild strawberries to the horns of their cattle as an offering to the elves. They believe that the elves, who are passionately fond of the berries, will help produce healthy calves and abundant milk in return.

Strawberries, first wild and then cultivated, are native to most temperate zones of the world, including regions in the Southern Hemisphere. The French royal family began the cultivation of strawberries in about 1300. Records dating from that time show that a great many varieties were already under cultivation in the royal garden. Modern cultivated varieties date from about 1714, when a French spy named Captain Frezier took a strain from Chile to France. In France the Chilean berry was crossed with a North American import that had been grown in North America since 1624. From this cross developed the myriad varieties currently available. Active breeding in this country, however, did not begin until the late 1890s.

Many medicinal uses were claimed for the wild fruit and its leaves. For example, it was believed that the berries alleviated symptoms of melancholy, fainting, all inflammations, fevers, throat infections, kidney stones, halitosis, attacks of gout, and diseases of the blood, liver, and spleen. Strawberries were taken to help shed "excess flesh," and the roots, chewed, removed tartar from teeth. A beverage brewed from the roots salved the pains of ulcers. Other strawberry preparations bleached out freckles and whitened skin, cured jaundice, and assuaged heart palpitations. Soaked in wine, the leaves and roots soothed the liver; soaked in water, the eyes. The same potion tightened loose teeth, and even was thought to cure venereal disease. Tea brewed from the leaves was said to be a kidney specific.

The wide distribution of wild strawberries is largely from seeds sown by birds. As they eat the berries, birds swallow the seeds and pass them intact through their digestive tracts. The seeds sprout easily wherever deposited in reasonably suitable conditions. The germinating seeds respond to light rather than to darkness and moisture, and therefore need no covering of earth or any other material. This characteristic of strawberry seeds should be kept in mind should you ever attempt to grow strawberries from seed.

Although it demands more care than some fruits, the strawberry is not difficult to grow. Tidy sets of leaves, each composed of three toothed leaflets, spring from a central crown. White, slightly scented flowers grow up from the same central crown, and stand above surrounding leaves. Runners reach out in every direction, rooting willingly of their own accord, and then proceed to blossom and bear fruit as prolifically as the mother plant. The plants make lovely edgings or borders for flower beds, vegetable gardens, or walkways. Certain varieties (particularly everbearers) are also suited for ornamental planting in hanging baskets, jars, barrels, window boxes, pyramids, and for training against walls and trellises.

However, trellising is a risky proposition, especially in areas where winters are harsh, for trellised plants are extremely tender. The variety Lakeland lends itself to trellising better than most because of its hardiness and its growing habits. Advertisements for strawberries grown from seed, except a specific type of European wild strawberry, are simply a come-on. Sprouting the plants from seed in nursery or garden is extremely difficult, in contrast to conditions in the wild. The rate of germination tends to be poor, the rate of seedling survival is low, and plants sprouted from seeds will take an additional year beyond the normal two to begin producing fruit. Experimentation with strawberry seeds can be enthralling, but it is best not to count on this method for a full-fledged bed of plants.

Actual production of berries does not begin until the second season the plants are in the ground (the third for those grown from seed). The first season, except for everbearing varieties, all blossoms should be pinched off. This lets the plant concentrate all its energy on establishing a strong root system and a healthy set of leaves to support the following year's berry crop. However, everbearers set out in the spring, their blossoms pinched until about mid-July, will produce a modest crop that same fall, followed by a bumper crop next season. Most varieties taper off in their production the third season, becoming less and less productive each successive year. For maximum yield, therefore, it is best to plant a new bed every other year,

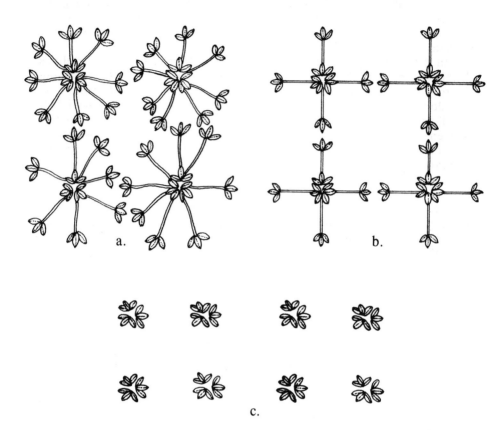

a.

b.

c.

Planting patterns for strawberries.
(a). *Matted row: space plants 18-24 inches apart in rows 2-3 feet apart and allow runners to root freely.*
(b). *Spaced matted row: set plants 18-24 inches apart in rows 2-3 feet apart. Pinch off all but the strongest runners — from four to eight — and set them evenly around the mother plant.*
(c). *Hill: set plants 12-18 inches apart in double or triple rows, separating sets of rows by 2-3 feet. Pinch off all runners. Best for everbearers or other varieties that produce few runners.*

using two or more sites, so that one group of plants is always in production, another getting itself ready.

For best results from strawberries, use one of three growing arrangements: the hill system, which requires the most care, but yields the biggest berries; the matted row system, which takes the least attention, but yields small berries, usually in profusion; and the spaced matted row system, intermediate in terms of both care and yield.

In the matted row system, plants are set out 1½ to 2 feet apart, in either double or single rows spaced 2 to 3 feet apart. If set out in double rows, the two rows of each pair are likewise 1½ to 2 feet apart, and 2 to 3 feet from the next double set. The first year they are set out, all blossoms are pinched off but all runners are allowed to root. Runners first appear as long leafed stems. At a length of about 8 inches they make a right-angle turn and send out roots. This unrestrained production and rooting of runners results in dense matting along the rows, from which the system gets its name. The aisle between the rows, or pairs of rows if such is the case, should be kept clear so berries can be reached, and plants cultivated easily.

The spaced matted row system is identical to the matted row system in details of spacing and basic arrangement of plants. However, in this system, only the few strongest early runners of each plant are allowed to root. And those that are allowed to root are carefully arranged around the mother plant so that each runner will have ample room to grow. Runners left from any one plant are limited to 4 to 8, depending on space available. Again, keep the lane between rows clear. With this system, renewal of exhausted beds is relatively easy. If the runners are correctly lined up, a cultivator run down the center of the row will plow under the tired mother plants, leaving the vigorous offspring to produce the next year's crop. This practice cannot be continued indefinitely, however, as it increases the risk of virus infection.

Under the hill system, plants are set at intervals of 12 to 18 inches, in double or triple rows. All runners are pruned off at roughly two-week intervals. Thus only the original plants grow and produce fruit; the berries they yield tend to be very large in size though fewer in number than under the two preceding methods. This system is best for plants that by nature tend to produce few runners, such as everbearing varieties. However, it can also be used for standard varieties where size rather than quantity is the grower's goal.

Actual planting must be done carefully, since setting plants in too deep or not deep enough will kill them. Soil level should exactly match that at which plants grew in the nursery. This means the level should hit right at the joint between roots and crown, right at the nursery mark. If the roots are exposed, they will dry out and the plant will die. Take time to make sure that plants are properly set in.

Strawberry plants are not fertilized their second spring, although the mulch should be replenished as needed. Fertilizer administered then would only encourage excess foliage, which would interfere with berry production.

If beds are left to grow a third season, treat as second-year beds, but expect fewer berries. Renewal of an aging strawberry bed is possible after the second or third season. Renewing spaced matted row plantings has already been described (page 53). However, beds planted by the matted row and hill systems can also be rejuvenated. Keep weeds at a minimum during the productive life of the bed. As soon as the harvest is finished the second or third summer, plow up the aisles between rows, restoring the rows themselves to their original 12- to 15-inch width. Next, set a power mower at a height of 3 to 5 inches, and mow off the plants. Fertilize with compost or rotted manure or balanced fertilizer, water thoroughly, and proceed as directed for any new bed of strawberries. This means pinch off blossoms the next spring, thin or eradicate runners, cultivate for weed control, and mulch well into the fall. The following season should see vigorous growth and production from the renewed patch.

HARVEST

Strawberries taste best if picked when fully red and ripe, and should be processed right away. The crop comes along quickly once it has begun to mature, and therefore picking should be done every day, or at least every other day. Pick in the morning after dew has dried, or in the cool of late afternoon and evening. Berries picked late in the day can then go straight on a shortcake or into a bowl of cream.

When picking, be sure to take the hull and a small piece of the stem as well as the berry. Then remove the green hull when it is time to process the berries. This keeps the fruit firm and fresh. Wash the berries, if necessary, just before eating or processing. By following both of these procedures, taste and appearance of the berries will be much enhanced. If the hull is left behind or the berries are washed long before consumption, the end product will look and taste decidedly half-hearted.

Wild varieties, whether from Europe or North America, continue to be popular with many gardeners. Their flavor is unique, and those who grow them swear there is nothing remotely comparable among cultivated kinds. Plants, blossoms, and berries are smaller, more delicate than their cultivated cousins, and some wild strains such as Alpine or Charles V or Catherine the Great produce no runners. Therefore, they make excellent plants to use ornamentally, especially in borders.

PROBLEMS

With strawberries as with other berries, diseases are best prevented by proper culture of the plants. Most devastating of the diseases that attack strawberries are the funguses, especially verticillium wilt, red stele, and gray mold. Not all areas of the country are prone to these funguses, so check with the local agricultural extension office and/or a knowledgable nurseryman before making your choice of varieties.

Verticillium wilt attacks plants at the peak of their first season, and causes them to wilt and collapse. The fungus gradually builds up in the soil, so avoid growing susceptible plants in the same soil. Rotate strawberries off the bed after three or four years and let the bed rest for three or four years.

Gray mold, or berry rot, attacks bloom and fruit, and causes both to rot. Like most fungus diseases, gray mold is most likely to appear during wet seasons.

Red stele is the most prevalent strawberry disease in the Northeast. It is borne in the soil and attacks the root core, causing it to turn reddish and mushy. Soil that has become infected with the fungus is practically impossible to purify, and must be abandoned for growing strawberries and all other stele-susceptible crops.

Black root rot is actually several different diseases, all with the same effect on the plant: small feeder roots are killed and the main root system is then invaded. It turns black and compromises the health of the plant. Controls include rotation of crops on a regular basis, selection of a good planting site, and use of strong healthy plants to begin with.

Leaf spot and leaf scorch, both foliage diseases, are the result of a fungus that infects the parts of the plant above the ground. Hot, wet weather promotes their occurrence, especially during August and early September. Fungicides available at the local garden supply center can be used to control these infections, but check with the county extension service for specific recommendations, and then follow directions on fungicide package carefully.

Viruses that afflict strawberries are numerous and almost unavoidable. They can cause the leaves to turn yellow and the plants to decline and stop growing, and are responsible for gradual loss of vigor in plants kept for several seasons. Plant only certified virus-free plants, and control aphids.

Nematodes, present in the soil where crops have been grown intensively, feed on the microscopic roots of the strawberry plants, leaving wounds that permit fungus and virus diseases to enter the plant. Control of nematodes, therefore, is important in warding off other diseases. Rotating crops regularly is one easy way to cut down on nematode population. Soil fumigation also helps, but is impractical on other than a large-scale commercial basis.

Insects that make the strawberry the food of choice are almost countless. Most infestations are not completely debilitating, but each will leave its mark. Most destructive of the insects is the strawberry weevil, or strawberry clipper. It is particularly prevalent in the Northeast, but not unknown elsewhere. Adults cut off fruit buds early in the blossoming period: the female lays her eggs inside the bud, then girdles the stem and leaves the bud either hanging or fallen. The larvae hatch inside the bud, emerging in early summer as adults that feed on the pollen of the blossoms. Several spraying compounds are effective; consult your local agent.

Grubs, wireworms, and cutworms are a problem in soil that has recently been in sod. They delight in the tender root systems of young strawberries, and can thoroughly weaken the plants. Thorough cultivation of newly-turned land and a year's lapse in planting strawberries in it should prove adequate controls.

Cyclamen mites feast on the tender young leaves as they emerge from the center of the crown. If they are present in numbers, they cause stunted growth and faded leaves. Use of ladybugs and praying mantises is a good natural control, although in serious situations pesticides may be required as well. Check with your local agent.

TABLE 9.

GROWER'S GUIDE TO STRAWBERRIES

A. Selecting a Variety. To keep supply constant, choose at least 2 varieties with different ripening times — perhaps a June-bearer and an everbearer. June-bearers give one large crop; everbearers produce a light crop during early and midsummer, and a larger one in late summer.

B. Site. Slight elevation, gentle slope, or raised beds; well-drained in full sun.

C. Soil. Loamy and rich in organic matter; pH 5.8-6.5 (but will tolerate 5.5 or 7.5).

D. Planting. Plants come bare-rooted in batches of 25 (usually). Soak roots for 1-2 hours in water. Plant in early spring, after all danger of frost is past and soil is workable. (In the South, also plant in fall or as late as January.) Set each plant into a hole 6 inches deep, spreading out the roots. Fill in soil around roots and plant so that soil level, tamped down, comes *exactly* to the nursery mark (joint between roots and crown). Water and refirm the soil around the plant to the correct level. Water again every 3-4 days until plant is established.

E. Planting Patterns.
  Matted Row System: set plants 18-24 inches apart in rows 2-3 feet apart.
  Spaced Matted Row System: set plants 18-24 inches apart in rows 2-3 feet apart.
  Hill System: set plants 12-18 inches apart in rows 2-3 feet apart.

F. Cultivating. To keep weeds down during first half of plant's first season, work soil regularly to depth of 2-3 inches around plant, taking care not to injure plant or runners. Then mulch (see below).

G. Mulching. Four to six inches deep around plants with straw, grass clippings, wood chips, rotted sawdust, leaves, or pine needles. When temperatures drop below 20° F. in fall, cover the plants entirely with mulch to protect them from frost. In spring, after all danger of frost is past, uncover the plants and rearrange the mulch alongside the plants, replenishing as needed for a depth of 4-6 inches.

H. Fertilizing. Fertilize late June to early July the first year *only,* with side-dressing of 10-10-10 applied well away from stems and crowns and worked well into the top 2-3 inches of soil. Do not fertilize the second or third years except to renew beds (see Table 10).

I. Pruning.
  Blossoms. First year, pinch off *all* blossoms from June-bearing varieties; pinch off all blossoms *until* late June-early July from everbearers, letting later blossoms set and ripen. Second and third years, allow all blossoms to set, mature, and ripen.
  Runners. First, second, and third years: *Matted Row System* — allow all runners to root; keep aisles clear. *Spaced Matted Row System* — pinch off all but 4-8 of the strongest for each plant, fanning runners evenly around mother plant and sinking their roots in soil. A forked twig or hairpin will help hold the runners in place until firmly rooted. Keep aisles clear. *Hill System* — pinch off all runners; keep aisles clear.

J. Propagating. With wild varieties, dig up and divide the crown every third year. Standard varieties usually reproduce readily on their own. Runners that are the mature plant's offspring start growing just a few weeks after spring growth has begun, and those that are allowed to take root become part of the productive bed. They are also crucial to processes of renewing spent spaced-matted-row beds, as outlined in Table 10. Generally, everbearing varieties put out fewer runners than spring-bearing varieties. If these everbearers are kept free of runners, they will last six years or more. Plants and berries become very large. To refurbish a fading hill-system bed, replace plants with new stock.

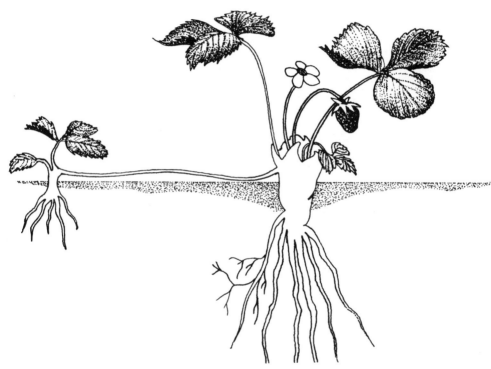

Cross-section of strawberry plant.
*At right is larger mother plant, with blossom, berry, leaves, and root system.
Crown of plant is at base of growth. At left is smaller runner plant, with
leaves, crown, and roots of its own. Runner sent out by mother plant connects
the two, and has formed daughter by sending out roots and taking up resi-
dence about 8 inches from mother.*

Two-spotted mites, red mites, and/or spider mites feed on the backsides of leaves, and make them warped and pale. The mites are most often present during abnormally dry seasons. Spray affected plants with water to wash off the mites and the dust that encourages them.

Root weevils are the larvae of other insects. The weevils feed on the roots and crowns of strawberry plants, causing weakness and stunted growth. The most effective control dictates plowing under a heavily-infested patch right after harvest.

Crownborers do just what their name implies: eat into the crown of the plant. Plow under the afflicted crop as soon as harvest is over, and establish any new bed at least 300 feet away from the infested site. This distance seems to exceed the limit of the adult borer's mobility, and should prevent their spread.

The tarnished plant bug, a small bronze-colored beetle with white, yellow, and black blotches, is rapidly becoming more of a problem. Its season extends over most of the warm summer months, and its effects are seen in dwarfed, button-shaped berries. In order to control the beetle, it is often necessary to use a spray that leaves behind a residue. Check with your local agent.

Spittlebugs can be detected by the pockets of spittle they leave on strawberry leaves. To combat this pest, follow the county agent's advice on spraying.

Aphids come winged and wingless, yellowish, greenish, or green and black. Not only do they carry viruses, but they eat the new growth at the center of the crown, and must be controlled if vigorous beds are to be maintained. The cornfield ant or ladybug can be used as a natural control, or a commercially-prepared spray will serve the same purpose. Check with your local agent. Controls effective against aphids will also deter the leaf-sucking leafhoppers.

VARIETIES

There are two types of strawberry varieties: June-bearing and everbearing. The June-bearers bear during the first half of the summer, while the everbearers produce a light crop during early and midsummer, and then produce a much larger crop in late summer. June-bearers are further divided according to date of ripening: early, mid-season, or late. Both types are available in a wide assortment of varieties specifically suited to different areas of the country.

The following list includes area of suitability, ripening date, and other important data. The roster is far from exhaustive, listing the major varieties only, although a few unusual varieties are included at the very end. If not specifically defined as an everbearer, the variety is June-bearing.

Albritton — for mid-South, ripens late; freezes well, tastes good.

Atlas — North Carolina to just north of the Gulf Coast; ripens in mid-season; susceptible to red stele and powdery mildew, but resistant to leaf scorch.

TABLE 10.

RENEWING OLD STRAWBERRY BEDS

A. Matted Row/Hill Systems. In late fall, plow up aisles between rows and mow off plants at 3-5 inches. Mulch for winter. In spring, pull back mulch, fertilize, water, and treat as first-year beds.

B. Spaced Matted Row System. In late fall, plow under mother plants, saving runners to become the new mother plants. Mulch for winter. In spring, pull back mulch, fertilize, water, and treat as first-year beds.

C. Hill System Everbearers. Kept free of runners, plants will last about 6 years. When vigor declines (smaller crops and fewer leaves), plow plants under and set out new ones.

Blakemore — produces small berries that hull easily, on plants that send out many runners; grows well from Virginia to Georgia, west to Oklahoma and southern Missouri; excellent for preserving; resistant to verticillium wilt, leaf scorch, and leaf spots; ripens early.

Catskill — sends out many runners; does well from New England to New Jersey, west to southern Minnesota, but is susceptible to virus diseases and leaf spots; ripens mid-season.

Dabreak — particularly good for the South; excellent for desserts and preserves; ripens early; resists leaf spots.

Earlidawn — thrives from Maryland north to New England and west to Missouri; berries ripen early, have light skin and flesh; plants are somewhat resistant to leaf spots and leaf scorch, but susceptible to verticillium wilt; good fresh or frozen.

Fairfax — sends out few runners; grows from southern New England to Maryland, west to Kansas; resistant to leaf spots and leaf scorch, but susceptible to virus diseases; ripens early.

Fletcher — excellent flavor, good frozen; plants produce many runners; does particularly well in New York and New England; ripens mid-season.

Florida Ninety — bred for conditions specific to Florida; good fresh; sends out runners prolifically; highly susceptible to leaf spots and leaf scorch; ripens early.

Fresno — does especially well in southern California; bears light spring crop and heavy midsummer crop; resistant to viruses; should be planted in the summer.

Gem (also called Brilliant and Superfection) — good fresh; resistant to leaf scorch, but not to leaf spots; everbearer; does well in northern soils from Michigan eastward.

Guardian — adapted to the central East Coast westward to Missouri; resistant to red stele, verticillium wilt, leaf scorch, powdery mildew; susceptible to leaf spots.

Headliner — for southern states; good for desserts; resistant to leaf spots; ripens mid-season.

Holiday — does well in Northeast; good flavor, bears well; resistant to leaf spots and leaf scorch; ripens mid-season.

Hood — good for desserts and preserves; resistant to mildew, foliage diseases, and red stele, but susceptible to viruses; late ripening.

Jerseybelle — does well from southern New Jersey through northern states; does not freeze well, but produces large fruits; highly susceptible to leaf spots, leaf scorch, red stele, and verticillium wilt; ripens late.

Lakeland — everbearer, sends out unusually long runners that can be trained on a trellis; produces big berries June to September.

Marlate — very tasty; resistant to leaf scorch, leaf spots, and mildew; susceptible to red stele and verticillium wilt; does best in mid-Atlantic states; ripens late.

Midland — good for eating fresh or frozen; resistant to leaf spots and leaf scorch, but sensitive to virus diseases; grown in numbers from southern New England to Virginia, west to Iowa and Kansas; ripens very early.

Midway — best eaten fresh, also good for freezing; resistant to some strains of red stele, but susceptible to leaf spots, leaf scorch, verticillium wilt; most widely grown variety in Michigan, also does well in other northeastern states and south to Maryland; ripens mid-season.

Northwest — good fresh or frozen; resistant to viruses but not to leaf spots; widely grown in Washington and Oregon; ripens late.

Ogallala — sweet and flavorful; freezes well; winter hardy; resistant to leaf spots and drought; everbearer which grows well from Mississippi River through Rocky Mountains.

Olympus — good for freezing; resistant to virus diseases and some strains of red stele.

Ozark Beauty — susceptible to variegation or June yellows; everbearer.

Pocahontas — good fresh and frozen; resistant to leaf scorch, particularly resistant to leaf spots; does well from southern New England south to mid-Virginia, west to Missouri; ripens early.

Quinault — large round berries, soft; good fresh but not frozen; susceptible to mildew; does best in West; everbearer.

Raritan — large berries; susceptible to red stele, verticillium wilt; slight drought resistance; grows in same region as Jerseybelle; ripens mid-season.

Redchief — good frozen; resistant to red stele, verticillium wilt, leaf scorch, mildew; slightly vulnerable to leaf spots; does best in East; ripens early.

Redstar — good fresh; resistant to virus diseases, leaf spots, leaf scorch; grows well from southern New England to Maryland, west to Missouri and Iowa; ripens late.

Rockhill (also called Wazata) — an old variety; grown in Minnesota, Iowa, Oregon, northern states; everbearer which can be propagated by dividing crown.

Sparkle (also called Paymaster) — good fresh or frozen; resistant to most strains of red stele, some strains of leaf spot; susceptible to viruses; does well throughout the Northeast and west to Wisconsin; ripens late.

Stoplight — very good flavor; resistant somewhat to leaf spots and leaf scorch; grows well in North Central and Plains states; ripens mid-season.

Sunrise — light red at maturity, excellent flavor; best eaten fresh; resistant to red stele, somewhat resistant to verticillium wilt, leaf scorch, mildew; susceptible to leaf spot; grows best in South Central states; ripens early.

Surecrop — good for desserts; does well spaced 6-9 inches apart, so useful for borders and edgings; resistant to red stele, verticillium wilt, leaf spots, leaf scorch, drought; Mid-Atlantic states; ripens early.

Tangi — bred for the Gulf Coast; resistant to leaf spots, leaf scorch, but susceptible to anthracnose and powdery mildew; ripens early.

Tennessee Beauty — good for desserts and freezing; resistant to leaf spots and leaf scorch, tolerant of assorted viruses; popular for flavor, firmness, color, and productivity, and widely grown from Kentucky to Missouri; resists drought less well than many other varieties, so needs irrigation; ripens late mid-season.

Tioga — retains shape and appearance for a long time after picking, so is popular for shipping; susceptible to leaf spots; leading California variety, can be planted summer and winter to produce in March or April; produces for two months; grows well in Florida also.

Totem — good for desserts and freezing; resistant to viruses, moderately hardy; bred for use in the Northwest; resistant somewhat to fruit rot, resistant to mildew and red stele; ripens late.

Trumpeter — excellent flavor; freezes well, winter hardy, susceptible to leaf spots; grows well in Upper Mississippi region, Plains states; ripens late.

Unusual Varieties

Fraises des Bois — French (Charles V) or Russian (Catherine the Great) strains; neither puts out runners, but both can be propagated after two years by digging, dividing, and replanting the crown; excellent flavor; terrific for edging plants; bloom and fruit over long period; hardy, but need winter protection; keep moist.

Alexandria-Alpine — grows only from seed, readily reseeds itself; can be used as an ornamental or in jars or barrels; does not produce runners; fruit is small, tasty; plant is descended from wild varieties.

Wild Scarlet Strawberry (*Fragaria Virginiana*) — indigenous North American wild variety with small, flavorful berries; produces some runners.

Paris Spectacular — bred for hanging pots; produces fruit over extended period on main plant and unrooted runners.

Snow King — white strawberry; puts out runners prolifically, so suitable for use in hanging pots, barrels, jars; berries small, resemble wild varieties in taste.

GROWING AND COOKING BERRIES

COOKING BERRIES

CANNING & FREEZING

CANNING SOFT BERRIES — RASPBERRIES, BLACK RASPBERRIES, BLUEBERRIES

Freezing is preferable to canning, generally speaking, because it preserves the taste and color better. But canning is useful, too, and certainly fills the bill when freezing is impossible.

Light syrup:
1 cup sugar or honey
1 quart water

Medium syrup:
2 cups sugar or honey
1 quart water

Heavy syrup:
3 cups sugar or honey
1 quart water

Put sugar or honey and water into saucepan and stir to combine. Bring to boil and boil 5 to 10 minutes, until mixture thickens slightly. Put berries into sterile jars, pour boiling syrup over them to fill jars, and seal. Process in boiling water to cover jars for 20 minutes. Cool to room temperature in water bath, then store. A piece of cinnamon stuck into the jar before processing varies the flavor. Canned berries can be used to make pie, jam, or any other confection.

FREEZING BERRIES — SOFT OR FIRM

There are two methods of freezing berries. With the first method, the berries keep their shape; with the second, they do not, but become sweet and juicy.

1. Stem or clean fruit as necessary. Place gently in containers and freeze whole, without sugar.

2. Stem or clean fruit as necessary. Fruit may also be sliced or crushed. Add 1 cup sugar to each quart of fruit. Spoon into containers and freeze.

CANNING BERRY JUICE

Extract juice from any type berry by pureeing and then straining fruit. Put the juice, unsweetened, into sterile jars. Process 20 minutes in boiling water, cool in water, and store. Makes good jelly or spiced drinks in the middle of winter.

CANNING FIRM BERRIES — ELDERBERRIES OR BLACKBERRIES*

Again, freezing is preferable, but not always feasible.

1 quart berries   ½ cup water
½ cup sugar or honey

Combine in pan and heat slowly, stirring gently to avoid damaging berries. Simmer berries 5 to 10 minutes, then remove from heat and set aside for 2 hours. Return to boiling, pack in hot sterile jars, and seal.

*Strawberries tend to disintegrate, fade, and become unpleasantly mushy when canned. Therefore, either make into jam or preserves, or freeze.

BREADS

BLUEBERRY GRAIN MUFFINS

Extra-hearty muffins for a cold winter's morn.

⅔ cup milk
½ cup cornmeal
½ cup cooked rice
½ cup flour
2 tablespoons sugar
3 teaspoons baking powder
½ teaspoon salt
1 egg yolk, well beaten
1 tablespoon shortening, melted
1 egg white, beaten stiff
¾ cup blueberries, frozen or canned, drained

Preheat oven to 400° F. Scald milk and pour slowly onto cornmeal. Let sit 5 minutes, and then add rice. Stir to mix. Sift together dry ingredients and add to cornmeal mixture with egg yolk and shortening. Stir gently to blend. Fold in beaten egg white, then blueberries, and spoon into well-greased muffin pans. Bake about 25 minutes, until nicely browned. Serve warm. Makes 10-12.

BUTTERMILK BLACKBERRY BREAD

This also makes up well with blueberries, black raspberries, or elderberries.

1 cup white sugar
1 cup brown sugar
5 cups flour
1 teaspoon baking soda
5 teaspoons baking powder
1 teaspoon salt
2 eggs
2 tablespoons melted butter
2 cups buttermilk or sour milk
2 cups blackberries
1 cup coarsely chopped pecans or walnuts

Preheat oven to 350° F. Sift together dry ingredients. Beat eggs and combine with melted butter and buttermilk or sour milk. Stir into dry ingredients until well blended. Gently fold in berries and nuts. Pour into 2 greased 9x5-inch loaf pans. Let sit for 20 minutes before baking. Bake 1 to 1¼ hours, until firm and lightly browned. Makes 2 loaves.

ELDERBERRY BLOSSOM FRITTERS

Pour maple syrup or elderberry syrup, made from frozen elderberries, over these. Unbelievable.

1 quart elderberry blossoms, with stems
1 cup flour
¼ cup sugar
1 teaspoon baking powder

¼ teaspoon salt
1 egg
1 cup milk
Oil or fat for deep frying

Wash and gently dry blossoms. Sift dry ingredients together. Beat together egg and milk and beat into dry ingredients. Heat oil 1-inch deep in skillet. Dip blossoms into batter, holding them by stems, and drop into oil. Fry until golden. Drain and serve hot with syrup. Serves 8.

BAKED BLUEBERRY PASTIES

Tuck these in lunches for dessert, or serve without sauce for after-school snacks.

2 cups flour
2½ teaspoons baking powder
½ teaspoon salt
¼ cup shortening
¾ cup milk

1½ cups blueberries
½ cup brown sugar
¼ teaspoon nutmeg
2 tablespoons butter

Preheat oven to 350° F. Sift flour, baking powder, and salt. Cut in shortening. Pour milk into well in center, and stir until blended. Roll out to ¼-inch thickness and cut into 3-inch squares. Put blueberries in center of each square, sprinkle with sugar and nutmeg, and dot with butter. Bring corners of squares together at top and pinch seams. Place pasties in greased baking dish. Prick each pasty with a fork and bake about 45 minutes, until golden brown. Makes 6 to 8 pasties. Serve topped with Hard Sauce.

Hard Sauce:
4 tablespoons butter
1 cup confectioners' sugar
⅛ teaspoon salt

1 tablespoon heavy cream
1 tablespoon rum

Cream butter, gradually beat in sugar and salt until mixture is light and fluffy. Beat in cream and rum. Makes about ¾ cup.

RASPBERRY NUT BREAD

Substituting rhubarb for half the raspberries makes this extra special.

6 tablespoons butter
½ cup white or brown sugar
2 eggs
2 cups flour
1 teaspoon baking powder
½ teaspoon baking soda
½ teaspoon salt

¼ cup sour cream
1 cup raspberries, fresh or frozen,
   thawed and drained
½ cup coarsely chopped blanched
   almonds
1 tablespoon flour

Preheat oven to 350° F. Cream butter and sugar until light and fluffy. Beat in eggs, one at a time. Sift dry ingredients together and add to egg mixture alternately with sour cream. Toss together berries and nuts with 1 tablespoon flour and fold gently into batter. Pour into greased 9x5-inch loaf pan, or 2 small loaf pans. Bake 45 to 50 minutes, or until firm and lightly browned. Cool in pan 10 minutes, then turn out onto rack. Makes 1 large or 2 small loaves.

RASPBERRY MUFFINS

Muffins with a difference. Any berry works its own brand of magic.

1¾ cups flour
½ teaspoon salt
½ cup sugar
1 teaspoon baking powder
1 teaspoon vanilla
1 egg, well beaten

1 cup buttermilk
6 tablespoons (¾ stick) shortening,
   melted
½ cup raspberries
1 tablespoon grated orange rind

Preheat oven to 400° F. Sift together dry ingredients. Combine liquid ingredients in separate container and stir into flour mixture until barely blended. Combine raspberries and orange rind, and fold into batter. Spoon batter into 10 or 12 well-greased muffin cups. Bake about 25 minutes, until nicely browned. Turn out on rack. Serve warm. Makes 10 to 12.

BLUEBERRY ORANGE BREAD

Makes a good coffee cake or tea treat.

2 tablespoons butter
¼ cup boiling water
½ cup orange juice
1 tablespoon grated orange rind
1 egg
1 cup brown sugar
2 cups flour
1 teaspoon baking powder

¼ teaspoon baking soda
½ teaspoon salt
1 cup blueberries

Glaze:
2 tablespoons lemon juice
2 tablespoons honey
1 tablespoon grated lemon rind

Preheat oven to 325° F. Melt butter in boiling water and stir in orange juice and rind. Beat egg and sugar until light and lemon-colored. Sift together dry ingredients and add alternately to egg mixture with orange juice mixture. Beat until smooth. Gently fold in berries. Place in greased 9x5-inch loaf pan, or in 2 smaller loaf pans. Bake about 1 hour and 10 minutes, or until firm and golden. Take from oven and turn out on wire rack. Combine glaze ingredients and spoon over warm loaf. Cool before slicing. Makes 1 large or 2 small loaves.

BLACK RASPBERRY LOAF

A touch of mace and lemon livens this up nicely. Serve for breakfast or for tea; it's very rich.

4 whole eggs
4 egg yolks
1 cup sugar, white or brown
1¼ cups flour
¼ teaspoon salt
2 tablespoons cornstarch
¼ teaspoon mace

1 cup butter, melted and cooled
1 teaspoon vanilla
1 teaspoon grated lemon rind
1 cup black raspberries, tossed with 1
  tablespoon flour
½ cup crushed blanched almonds

Preheat oven to 350° F. Combine eggs, egg yolks, and sugar in bowl over hot water. Stir until mixture becomes warm and sugar has dissolved. Beat at high speed for 10 minutes, until thick and triple in volume. In another bowl, sift together dry ingredients and fold into egg mixture gradually. Combine melted butter and flavorings and stir into batter. Fold in floured black raspberries. Grease two 9x5-inch loaf pans and sprinkle bottoms with crushed blanched almonds. Pour in batter and bake 45 to 50 minutes, until firm and lightly browned. Cool loaves in pans for 10 minutes, then turn out on rack to finish cooling. When cool, dust tops with confectioners' sugar, slice, and serve with butter. Makes 2 large loaves.

BLUEBERRY PANCAKES

Traditional and delightful.

1½ cups sifted flour
2½ teaspoons baking powder
3 tablespoons sugar
¾ teaspoon salt
2 egg yolks

1 cup milk
3 tablespoons melted shortening
1 cup fresh blueberries
2 egg whites, beaten until stiff

Sift together dry ingredients. Beat yolks until light yellow and beat in milk and melted shortening. Stir into dry ingredients until mixture is smooth. Stir in blueberries, then gently fold in beaten egg whites. Bake on hot griddle. Serve with syrup. Makes about 8 pancakes.

ELDERBERRY BREAD DROPS

Better than biscuits and easier than bread.

2 cups flour
1 teaspoon baking soda
2 teaspoons cream of tartar
2 tablespoons sugar

⅛ teaspoon salt
1 cup milk
1 tablespoon butter, melted
2 cups stemmed elderberries

Preheat oven to 375° F. Sift together dry ingredients. Combine milk and butter and stir into dry ingredients. Lightly stir in elderberries. Drop by large spoonfuls on greased cookie sheet. Bake about 20 minutes, until lightly browned. Makes about 12.

DRINKS

RASPBERRY FRUIT PUNCH

Float fresh berries, slices of lemon and orange, and sprigs of fresh mint in this punch to give it a truly festive appearance. And to make it a "punch with punch" substitute white wine for part or all of the carbonated water.

4 cups sugar
4 cups water
2 quarts red raspberries, fresh or frozen, unsweetened
1 cup sliced pineapple, fresh or canned
1 cup pineapple juice
1 cup orange juice
1 cup lemon juice
3 sliced bananas
2 quarts carbonated water or white wine, or 1 quart of each
3 to 4 cups crushed ice

Combine sugar and water and boil for 10 minutes. Cool. Crush raspberries and force through sieve. Add to pineapple, juices, and bananas. Pour in cooled syrup and chill 3 to 4 hours. Immediately before serving add carbonated water and/or wine and crushed ice. Garnish with fresh berries, fruit slices, and mint. Makes about 5 quarts.

RASPBERRY SHRUB

A special zip makes this light drink a real pick-me-up.

1 quart raspberries, fresh or frozen, unsweetened
1 cup white vinegar
1 cup water
3 cups sugar

Combine raspberries, vinegar, and water in large bowl. Cover and let stand 24 hours. Squeeze through cheesecloth or put through sieve. To each 3 cups of strained juice, add 3 cups sugar. Cook juice and sugar, boiling, for 20 minutes. Cool and freeze until ready to use. To mix up, stir about ¼ cup into 8 ounces of water or fruit juice and pour over cracked ice. Makes 1 quart concentrate.

BLACKBERRY, BLACK RASPBERRY, OR ELDERBERRY SHRUB

Substitute 1 quart of one of these berries for the red raspberries in the preceding recipe, and proceed as directed.

FRUIT PUNCH

Perfect for an afternoon party. The tea acts as a catalyst to blend all the flavors.

1¼ cups sugar
1¼ cups water
2½ cups strong, hot tea
2½ cups strawberry juice, from 1
    quart crushed strawberries

1 cup crushed pineapple
1 cup lemon juice
1½ cups orange juice
1 cup maraschino cherries, drained
1 quart carbonated water

Combine sugar and water and boil for 10 minutes. Stir in hot tea and then cool mixture thoroughly. Crush and strain strawberries, measure juice, and add with pineapple, lemon juice, and orange juice to sugar mixture. Chill for 1 to 2 hours. At serving time, add water to make 4 quarts liquid. Stir in maraschino cherries and carbonated water. Pour into punch bowl and adorn with chunks of frozen strawberry juice. Makes 5 quarts punch.

FRESH BERRY MILK SHAKE

Thick, rich, and frosty.

1 cup milk
1 cup vanilla ice cream

1 cup berries

Combine ingredients in blender and blend until thick and smooth. Makes about 2½ cups. Use different berries, proceeding as above, and concoct milk shakes of entirely different flavors.

BERRIED ICED TEA

Nothing beats this on a hot humid day in berry season. Good with any of the berries in this book.

⅓ cup berries, crushed
¾ cup strong tea
2 tablespoons granulated sugar

2 tablespoons lemon juice
Ice cubes, or cubes of frozen juice

Puree fruit in blender and strain to remove seed. Pour into tall glass. Add tea, sugar, and lemon juice and stir to dissolve sugar. Plunk in ice cubes or juice cubes, garnish with fresh fruit or sprig of mint, and serve. Makes one glass.

PIES & TARTS

CRUST RECIPES

Each makes pastry sufficient for a 9-inch, 2-crust pie.

Rich Pastry:

¾ cup chilled butter   2 tablespoons sugar
2¼ cups all-purpose flour   1 egg

Combine chilled butter, flour, and sugar in bowl. Cut with two knives or pastry cutter until mixture resembles coarse meal. Beat in egg, and work mixture with fork or fingers until smooth. Shape into ball, wrap, refrigerate 1 hour or until firm. Then roll out on floured surface and proceed as recipe directs.

Plain Pastry:

½ cup shortening   ¼ teaspoon salt
2¼ cups flour   5 to 7 tablespoons ice water

Combine shortening, flour, and salt in bowl. Cut with two knives or pastry cutter until mixture resembles coarse meal. Pour in ice water and stir into mixture quickly. Gather dough into ball; if it crumbles, stir in 1 or 2 more tablespoonsful water. Then wrap in aluminum foil or plastic wrap and chill 1 hour or more before using. Proceed as recipe directs.

ANY BERRY PIE

A general rule that can be varied by the addition of 1 teaspoon cinnamon, ginger, or nutmeg here or there, or by the addition of ¼ to ½ cup berry vinegar.
Vary the taste by combining 2 or more kinds of berries for a special treat.

1 recipe plain pastry   2 tablespoons flour or 1 tablespoon
2½ cups berries, fresh, frozen, or   cornstarch
    canned   ⅛ teaspoon salt
½ cup sugar or less, depending on
    whether or not fruit is
    pre-sweetened

Preheat oven to 425° F. Roll out crust for 9-inch pie, and place bottom crust in pie pan. Place fruit in crust and sprinkle with sugar, salt, flour or cornstarch, and spice if used. Add top crust, seal and flute edges, and cut decorative slits in top. Brush with softened shortening or beaten egg yolk if desired, for a shiny crust. Bake about 40 minutes, until fruit is cooked and crust is nicely browned. Cool slightly and serve. Serves 6.

WHOLE EGG MERINGUE

This makes a simple and irresistible topping to use instead of a crust on any berry pie. If this meringue is used, don't add any sweetener to the fruit.

1 large egg
1 cup sugar

1 teaspoon vanilla extract

Combine ingredients and beat on high speed of electric mixer until thick and stiff. Spread evenly over fruit in pie crust. Bake about 40 minutes at 425° F. until nicely golden and crunchy-crisp.

STRAWBERRY RHUBARB TART

Traditional fare when the first strawberries ripen in the spring.

Filling:

1 cup chopped rhubarb
2 cups fresh hulled strawberries
1 cup sugar

½ cup water
1 recipe plain pastry

Combine rhubarb, 1 cup berries, sugar, and water in saucepan. Bring to boil over moderate heat, and simmer, stirring occasionally, for 30 minutes. Mixture should resemble thick puree. Cool and chill. Meanwhile, preheat oven to 400° F. and prepare 1 recipe plain pastry. Roll out pastry and line a 9-inch pie tin. Prick bottom of crust and crimp edges decoratively. Line crust with waxed paper, fill with uncooked rice, and bake 15 minutes. Remove rice and paper and bake 10 minutes more, until crust is golden. Place on rack to cool.

Custard:

4 egg yolks
⅓ cup sugar
⅛ teaspoon salt
1 cup scalded milk

1 envelope plain gelatin
3 tablespoons kirsch liqueur
1 teaspoon vanilla
1 cup whipping cream

Beat egg yolks, sugar, and salt until thick and lemon-colored. Pour scalded milk into mixture in a continuous stream, beating constantly. Cook mixture in top of double boiler over hot water, stirring, until custard thickens slightly and coats spoon. Soften gelatin in 2 tablespoons cold water and stir until gelatin dissolves. Stir in flavorings and cool custard until thick but not set. Whip cream until stiff and fold into custard.

To assemble pie, spread puree in shell, top with custard, and garnish with reserved whole berries. Chill 1 hour or longer before serving. Serves 6.

ALEXANDER TORTE

Alexander was called "The Great;" so is this.

1 recipe rich pastry, chilled   1½ cups raspberry jam or preserves
2 tablespoons softened butter

Preheat oven to 250° F. Roll out chilled pastry dough to form two 10x10-inch squares. Spread each square with butter, place on greased and floured baking sheet, and bake about 40 minutes, until golden. Make filling while pastry bakes. Force jam or preserves through sieve, then cook over moderate heat for 3 to 5 minutes, until mixture forms thin puree. Spread mixture on one square of pastry and top with second square.

Icing: ¼ cup cold water
2 cups confectioners' sugar   2 teaspoons lemon juice

Combine ingredients, stir until smooth, and spread over warm torte. Cool torte, cut it into squares, and serve. Makes 2 to 3 dozen squares.

BLACK RASPBERRY JAM AND SOUR CREAM PIE

Thick, rich, and heavenly.

1 recipe plain pastry   2 egg yolks
2 cups toasted ground almonds   ⅓ cup sour cream
1 tablespoon milk   2 teaspoons cinnamon
⅔ cup black raspberry jam   1 tablespoon softened butter

Preheat oven to 425° F. Soak almonds in milk 5 to 10 minutes. Force jam through sieve to remove some seeds. Beat egg yolks, sour cream, cinnamon, and nuts into jam. Roll out dough to make bottom and top crusts for 9-inch pie. Pour filling into bottom crust, add top crust, crimp and seal edges. Cut decorative slits in top and brush with softened butter. Bake 30 minutes or until golden. Cool to room temperature before serving, and serve topped with whipped cream. Serves 6.

BLACKBERRY ALMOND TARTS

Slightly crunchy on top, smooth and delicious inside.

1 recipe rich pastry   2 eggs, well beaten
½ cup blackberry jam   1 tablespoon flour
¼ cup butter   1 teaspoon almond extract
⅔ cup sugar   1 cup cream for whipping
⅛ teaspoon salt   ½ cup chopped toasted almonds

Preheat oven to 425° F. Roll out pastry and use it to line 8 greased muffin tins. In each cup put 1 tablespoon jam. Cream butter and sugar until light, then beat in salt, well-beaten eggs, and flour and flavoring. Fill each cup ⅔ full with mixture. Bake about 12 minutes, until crust and top of tarts are golden. Whip cream, spoon onto tarts, sprinkle with toasted almonds, and serve while warm. Serves 8.

DEVONSHIRE CREAM PIE

Sort of a tall tart, this is fairly oozing with creamy custard.

Custard:

2 cups scalded milk

2 egg yolks

¼ cup sugar

⅛ teaspoon salt

½ teaspoon vanilla

Lightly beat yolks in top of double boiler. Beat in sugar and salt. Slowly stir hot milk, a little at a time, into yolk mixture. Cook in double boiler over hot water, stirring constantly, until mixture becomes thick and lightly coats spoon. Do not overcook or custard will curdle. Cool and stir in vanilla.

Pastry and Berry Filling

1 recipe rich pastry

2 cups blackberries

½ cup sugar or honey

Preheat oven to 400° F. Roll out pastry and cut out three 9-inch circles. Place 2 on greased baking sheet. Cut out center from third, leaving a 1½-inch ring, and place on the greased baking sheet. Bake 20 to 25 minutes, until lightly browned.

Place 1 round on serving plate. Spread it liberally with custard filling. Place second round on top and spread it, too, with custard. Place ring on top of custard. Mash berries slightly and stir in sweetener. Spoon berries into ring. Top with whipped cream, if desired, or with softened ice cream. Serves 6.

GLACE STRAWBERRY PIE

The strawberries stand up fresh and tall, and take on an enticing gleam from the sauce. Other fresh berries make excellent substitutes.

1 baked pie shell (plain pastry)	¾ cup water
1 cup sugar	1 teaspoon lemon juice
3 tablespoons cornstarch	3 cups strawberries
¼ teaspoon salt	1 cup cream for whipping

Combine sugar, cornstarch, salt, water, and lemon juice in saucepan. Cook, stirring, over low heat until mixture thickens. Simmer in double boiler 20 minutes. Cool.

Place fruit, part of it sliced, in baked shell. Place whole berries on top. Pour cornstarch mixture over it. Chill. At serving time, whip cream and pile on top of pie. Serves 6.

CHOCOLATE RASPBERRY PIE

This may well be one of the devil's chief instruments of corruption. No one can resist it.

Crust:

1¼ cups graham cracker crumbs	¼ cup finely chopped pecans or
¼ cup softened butter	walnuts
	1 tablespoon sugar

Combine crust ingredients, stir well, and press against bottom and sides of greased 9-inch pie plate. Bake at 375° F. for 8 minutes and then cool on rack.

Filling:

½ cup softened butter	1 cup whipping cream
⅓ cup sugar	1 cup fresh raspberries, or one
2 ounces unsweetened chocolate,	10-ounce package frozen,
melted and cooled	thawed and drained
2 eggs	

Cream butter with sugar until light and fluffy. Stir in chocolate, and beat in eggs, one at a time. Beat mixture at high speed for 3 minutes. Whip cream until stiff and fold it into mixture. Fold in raspberries. Spoon filling into cracker crust and chill 2 to 3 hours.

Glaze:

⅓ cup currant jelly	2 tablespoons water
½ cup juice from berries	2 tablespoons kirsch or cassis
1 tablespoon cornstarch	

Heat jelly and raspberry juice in saucepan, stirring. Dissolve cornstarch in water and stir into mixture; bring to boil, and cook, stirring, until thick. Remove from heat and stir in kirsch or cassis. Cool to room temperature. Spoon over pie. Serves 6.

ANGEL PIE

This one is airy and light, but devastatingly sweet as well.

Crust:

4 egg whites	1 cup granulated sugar
¼ teaspoon salt	½ teaspoon vanilla
¼ teaspoon cream of tartar	½ cup whipping cream

Preheat oven to 275° F. Grease a 9-inch pie pan. Beat egg whites until stiff, then beat in salt and cream of tartar. Beat in sugar gradually, and add vanilla. Spread meringue in pan, making an indentation in center. Bake about 1 hour, until dry and firm but not too brown. Cool meringue in oven with door open. When cool, remove from oven. Whip ½ cup cream and spread on meringue. Chill in refrigerator 2 to 3 hours.

Filling:

2 cups strawberries	1 tablespoon rum or sherry
½ cup sugar	½ cup whipping cream

Slice strawberries and stir in sugar. Add flavoring. Spoon into meringue shell. Whip cream and pile it on top. Serves 6-8.

SOUR CREAM ELDERBERRY PIE

The rich tart taste of elderberries at their best.

1 recipe plain pastry	1 cup sour cream
2 cups fresh elderberries, or two	2 eggs
10-ounce packages frozen,	¼ cup sugar, white or brown
thawed	1 tablespoon flour
(¼ cup water)	

Preheat oven to 350° F. Roll out pastry and line a 9-inch pie tin, fluting edges to make a high rim. Place berries in a saucepan with water and cook gently 10 minutes. (If frozen berries are used, cook in their juices instead of water.) Pour into unbaked pie crust. Beat sour cream, eggs, sugar, and flour together until thoroughly blended. Pour over berries. Bake 40 to 45 minutes, until center is firm and top is lightly browned. Serve warm or cooled to room temperature. Serves 6-8.

MOUNTAIN JELLY PIE

1 baked pie shell (plain pastry)
⅓ cup raspberry or strawberry jelly
½ cup softened butter
⅔ cup sugar

3 egg yolks
¼ cup lemon juice
3 egg whites
⅓ cup sugar

Preheat oven to 450° F. Spread jelly in cooled shell. Cream butter and ⅔ cup sugar, then beat in yolks one at a time. Beat in lemon juice. Place in oven and bake 5 minutes. Turn heat down to 375° and bake 20 minutes more. Pie should be browned but not set. Remove from oven.

Preheat oven to 425° F. Beat egg whites until stiff, then gradually beat in ⅓ cup sugar. Continue beating until meringue is stiff and glossy. Heap on pie and bake in upper part of oven about 5 minutes, until meringue is hot and lightly browned. Serve pie warm. Serves 6.

RASPBERRY CHIFFON PIE

This chiffon pie is lighter than air, and sinfully good.

1 baked pie shell (plain pastry)
½ cup sugar
1½ cups raspberries
1 envelope unflavored gelatin
¼ cup sugar

¾ cup water
1 tablespoon lemon juice
⅛ teaspoon salt
2 egg whites, beaten stiff
1 cup whipping cream

In large bowl, stir sugar into raspberries and let stand 30 minutes so berries "juice up." In saucepan, combine gelatin, ¼ cup sugar, water, lemon juice, and salt. Cook, stirring, over low heat until gelatin dissolves. Stir mixture into berries and chill until slightly thickened but not set. Beat egg whites until stiff and fold into berry mixture. Spoon into pie shell. At serving time, whip cream and pile on top of pie. Garnish with a few whole raspberries or a sprig of fresh mint. Serves 6.

CAKES AND CAKE DESSERTS

RASPBERRY KUCHEN

Made with yeast, so its flavor and texture are distinctive.

1 cup scalded milk
¼ cup honey
⅓ cup shortening
½ teaspoon salt
1 cake or packet yeast
1 egg, well beaten
¼ teaspoon mace

3 cups all-purpose flour
2 cups fresh raspberries, or two 10-
ounce packages frozen, thawed
and drained
⅓ cup sugar
1 egg yolk
¼ cup cream

Pour scalded milk into large bowl. Stir in honey, shortening, and salt and cool until lukewarm. Stir in yeast and let stand 5 minutes. Stir in beaten egg, mace, and flour, blending thoroughly. Cover and set in warm place to rise until doubled in bulk, about 45 minutes. Preheat oven to 350° F. Spread dough on bottom and sides of greased 9-inch pie plate. Place raspberries on dough. Sprinkle with sugar and drizzle combined egg yolk and cream over top. Bake 30 to 35 minutes, until nicely browned. Serve warm, with cream or without.

ELDERBERRY STREUSEL CAKE

Grand served hot for breakfast, or later with coffee.

¼ cup shortening
¾ cup sugar
2 cups flour
2 teaspoons baking powder
⅛ teaspoon salt

2 eggs
½ cup milk
1 teaspoon vanilla
2 cups elderberries, lightly floured

Preheat oven to 325° F. Cream shortening and sugar until light and fluffy. Sift together dry ingredients. Beat eggs into creamed mixture, then beat in flour alternately with combined milk and vanilla. Fold in elderberries. Pour into greased 8- or 9-inch pan.

Topping:
½ cup sugar
¼ cup flour

½ teaspoon cinnamon
¼ cup shortening

Combine topping ingredients until crumbly, sprinkle on top of cake, and bake 25 to 30 minutes, until cake springs back lightly when touched. Serves 6-8.

STRAWBERRY CRUNCH

Crunchy and smooth, light and rich in alternate layers.

Crunch:

1 cup flour
2 cups brown sugar, firmly packed
½ cup shortening, softened

½ cup chopped nuts (pecans, walnuts, or almonds)

Preheat oven to 350° F. Combine flour, brown sugar, shortening, and nuts, and stir until coarsely blended. Spread in shallow greased pan and bake for 15 minutes. Stir mixture often while baking. Remove from oven and cool.

Filling:

2 cups fresh strawberries, or two 10-ounce packages frozen, partially thawed
¾ cup sugar (less if berries are already sweetened)

1 tablespoon lemon juice
2 egg whites
1 cup whipping cream

Puree strawberries by running through blender or forcing through sieve. Stir in sugar if berries are not already sweetened. Add lemon juice. Beat egg whites separately until stiff, and fold berry puree into them. Whip cream until stiff and fold into berry mixture.

Crumble crunch mixture. Spread half in bottom of 9x14-inch pan. Spoon berry fluff over crunch. Sprinkle second half of crunch on top. Freeze until almost firm and serve. Serves 8-10.

STRAWBERRY SHORTCAKE

The traditional, and many say the only, way to eat fresh strawberries.

2 cups flour
2 teaspoons baking powder
½ teaspoon salt
2 tablespoons sugar
4 tablespoons butter

¾ cup milk
1 quart fresh strawberries
½ cup sugar
1 cup whipping cream

Preheat oven to 425° F. Grease large baking sheet. Sift together flour, baking powder, salt, and 2 tablespoons sugar. Cut in butter with two forks or pastry blender. Add milk gradually until mixture forms soft dough. Don't stir any more than necessary. Turn dough out onto floured board and pat out to ½-inch thickness. Cut into 4-inch rounds or shape into two 9-inch rounds. Place small rounds on baking sheet, or place one 9-inch round on sheet, dot with butter, and top with second 9-inch round. Bake either size about 12 minutes. Meanwhile, set aside 12 to 15 big beautiful berries for garnish. Combine remaining berries with ½ cup sugar and crush. Split rounds, warm from oven, with fork. Fill with crushed strawberries and spoon more on top. Whip cream until stiff and spoon lavishly over shortcake. Garnish with reserved berries. Serves 6-8.

BLACK RASPBERRY BUCKLE

This dish is traditionally made with blueberries, but it's even tastier with black raspberries.

½ cup shortening
1 cup sugar
2 eggs
1 teaspoon vanilla
¼ teaspoon lemon juice
2 cups sifted flour

½ teaspoon salt
1 teaspoon baking powder
1 teaspoon baking soda
1 cup sour cream
2 cups black raspberries

Cream together shortening and sugar until light and fluffy. Beat in eggs, one at a time. Beat in vanilla and lemon juice. Sift together dry ingredients and add to creamed mixture alternately with sour cream. Grease an 11x14-inch pan and pour batter in, spreading it out evenly. Sprinkle black raspberries on top of batter.

Topping:
½ cup sugar, brown or white
½ cup flour

1 teaspoon cinnamon
¼ cup shortening

Thoroughly blend all topping ingredients and spread over batter. Bake 45 to 50 minutes, until cake tests done. Serve hot with whipped cream or vanilla ice cream. Serves 8-10.

ICE CREAM BLUEBERRY SLUMP

A quick and easy route to a real humdinger of a dessert.

1 cup vanilla ice cream, softened
1 cup flour
1 teaspoon baking powder

¼ teaspoon salt
1 teaspoon cinnamon
1 quart blueberries

Preheat oven to 375° F. Place ice cream in mixing bowl. Sift together dry ingredients and stir into ice cream. Spread berries in 1½- or 2-quart baking dish and pour batter over them, making sure it touches dish on all sides. Bake 40 to 45 minutes, until cake springs back when touched lightly. Serve with more ice cream. Serves 4-6.

BLUEBERRY PUDDING CAKE

A surprising combination of ingredients put together in a surprising way produces a pudding cake that is the best surprise of all.

2 cups blueberries
3 tablespoons shortening
¾ cup sugar
1 cup flour
1 teaspoon baking powder

¼ teaspoon salt
½ cup milk
1 cup sugar
1 tablespoon cornstarch
1 cup boiling water

Preheat oven to 350° F. Grease a 1½-quart baking dish and put berries in it. In mixing bowl, cream shortening and ¾ cup sugar until light and fluffy. Sift flour, baking powder, and salt and add to creamed mixture alternately with milk. Blend well. Pour batter over berries. Combine 1 cup sugar and cornstarch. Sprinkle evenly over batter. Pour boiling water evenly over all. Bake 50 to 60 minutes, until cake is light and golden. Serve warm with ice cream. Serves 4-6.

BLUEBERRY CRUNCH

Best made with freshly-picked blueberries, but frozen or canned blueberries do nicely in winter.

1 cup quick-cooking oatmeal
1 cup brown sugar, firmly packed
½ cup flour
½ cup powdered milk
½ teaspoon salt

½ teaspoon cinnamon
½ cup shortening
1½ cups fresh blueberries, or one 10-ounce package frozen, thawed

Preheat oven to 350° F. Combine oatmeal, brown sugar, flour, powdered milk, salt, and cinnamon and mix thoroughly. Add shortening and mix until coarsely blended. Grease a 9x9-inch pan and spread half of mixture in it. Spread blueberries on crumbs. Sprinkle remaining half of crumbs on top of berries. Bake 40 to 45 minutes, until hot and bubbling. Let cool slightly before serving, and serve topped with vanilla ice cream. Serves 6.

BLACKBERRY JAM CAKE

Try pouring half a cup of rum or brandy over this and serving it like a fruitcake.

1½ cups butter, softened
1 cup white sugar
1 cup brown sugar
6 eggs, separated
2 cups blackberry jam
4 cups flour

2 teaspoons *each* ground cloves,
   cinnamon, nutmeg, and allspice
2 teaspoons baking soda
1½ cups buttermilk
½ cup chopped dates
½ cup chopped walnuts

Preheat oven to 350° F. Cream butter and sugars until light and fluffy. Beat in egg yolks, then jam. Sift together flour and spices. Stir baking soda into buttermilk. Add flour mixture and buttermilk alternately to creamed mixture. Stir in dates and walnuts. Beat egg whites until stiff and fold into cake. Grease and flour a 10-inch tube pan. Pour batter into pan and bake 50 to 60 minutes, cool in pan 10 minutes, then invert on rack to finish cooling. Sprinkle with confectioners' sugar.

SPICED STRAWBERRY CAKE

Cake like this comes along once in a long, long while. Served hot topped with slowly melting shredded cheese, this will bring everyone running.

Cake:

¼ cup shortening
½ cup sugar
1 egg
1 cup flour
1 teaspoon baking powder
¼ teaspoon salt

⅓ cup milk
½ teaspoon vanilla
2 cups hulled and sliced fresh
   strawberries, or two 10-ounce
   packages frozen, thawed and
   drained

Preheat oven to 375° F. Cream shortening and sugar until light and fluffy. Beat in egg. Sift dry ingredients together and add alternately with combined milk and vanilla to creamed mixture. Grease a 9x9-inch baking pan, and pour batter into it. Spread berries over batter.

Topping:

¼ cup softened shortening
½ cup light brown sugar
⅓ cup flour
½ teaspoon cardamon

½ teaspoon allspice
½ teaspoon cinnamon
½ teaspoon mace
¼ teaspoon nutmeg

Combine topping ingredients and blend with fork until consistency is even. Spread over berries and cake and pat down gently. Bake 40 to 45 minutes, until cake tests done. Serve hot topped with shredded cheese or whipped cream. Serves 6-8.

YOGURT CAKE WITH RASPBERRY TOPPING

The secret formula for this cake is finally revealed, but the results are still magically delicious.

Cake:

1 cup plain yogurt
1½ cups confectioners' sugar
3 eggs
¼ cup melted butter

2 cups sifted flour
1 tablespoon grated orange or lemon rind
1 teaspoon baking powder

Preheat oven to 350° F. Grease and flour a 9x9-inch baking pan. In bowl, beat together yogurt and confectioners' sugar. Gradually beat in eggs, butter, flour, and grated rind and continue beating until mixture is smooth. Add baking powder, beating as lightly as possible. Pour batter into pan and bake 40 to 45 minutes.

Topping:

2½ cups granulated sugar
3½ cups water
1 tablespoon lemon juice

2 tablespoons pistachio nuts or almonds, blanched and ground
1 cup whipping cream
2 cups raspberries, crushed

While cake bakes, combine sugar, water, and lemon juice in saucepan. Boil, stirring, until sugar is dissolved. Simmer 10 minutes without stirring. Remove from heat but keep hot. When cake is done, remove from oven and cut into squares or diamonds. Pour hot syrup over cake a little at a time so all syrup is absorbed. Cover and cool several hours. Whip the cream. Sprinkle ground nuts over cake. Place pieces of cake on serving dish and top with whipped cream and raspberries. Serves 6-8.

PINEAPPLE-BLUEBERRY PUDDING CAKE

A little of this luscious dessert goes a long way. But expect plenty of requests for seconds.

3 cups fresh blueberries
20-ounce can crushed pineapple, with juice

1 box 2-layer white or yellow cake mix
½ cup shortening, melted
1 cup crushed pecans

Preheat oven to 350° F. Grease a deep baking dish, and in it combine blueberries and pineapple. Sprinkle cake mix evenly over fruit. Pour melted shortening over cake mix, and sprinkle nuts over all. Bake 45 to 50 minutes, or until top is lightly browned and springs back when touched lightly. Serve warm with cream. Serves 6-8.

ELDERBERRY PANDOWDY

There's actually nothing dowdy about this dish. It's one of the very best.

Filling:
3 cups elderberries
½ cup molasses or brown sugar
¼ teaspoon nutmeg

¼ teaspoon cinnamon
¼ teaspoon salt
1 tablespoon lemon juice

Preheat oven to 400° F. Grease a 1½- to 2-quart baking dish. Combine all filling ingredients in saucepan and cook, stirring over moderate heat for 10 minutes. Pour into baking dish.

Topping:
1½ cups flour
2 teaspoons baking powder
½ teaspoon salt
½ cup white sugar

1 egg, well beaten
½ cup milk
½ cup shortening, melted

Sift together dry ingredients. Combine egg, milk, and shortening and stir gently into flour mixture. Spoon over hot elderberries. Bake 20 to 25 minutes, until top is brown and crusty. Serve hot with cold cream. Serves 6-8.

BLACKBERRY ROLL-UP

This forms a beautiful spiral when sliced to serve.

Crust:

½ teaspoon salt
2 cups flour
⅔ cup shortening

2 tablespoons lemon juice
3 tablespoons cold water

Mix salt and flour. Cut in shortening with two knives or pastry blender. Stir in lemon juice and water until dough forms a ball. Add 1 or 2 tablespoons more water if necessary. Wrap dough tightly in aluminum foil or plastic wrap and chill 1 to 2 hours. When thoroughly chilled, roll out to form 9x14-inch rectangle.

Filling:

1 cup blackberries
½ cup sugar
¼ teaspoon salt
¼ teaspoon nutmeg

½ cup gingersnap crumbs
1 teaspoon grated lemon rind
½ cup walnuts, broken into pieces

Preheat oven to 425° F. Grease a baking sheet. Mix together all filling ingredients and spread evenly on rolled dough. Starting at one short edge, roll dough up to form cylinder. Place carefully on baking sheet, seam side down. Bake at 425° for 10 minutes, then turn down heat to 350° and bake 15 to 20 minutes longer. Slice and serve hot with ice cream or cream. Serves 6-8.

BLACKBERRY UPSIDE-DOWN CAKE

Topsy-turvy treat, dripping with sweetness.

2 tablespoons butter
¼ cup brown sugar
2 cups blackberries
1½ cups white sugar
½ cup shortening
2 eggs

1½ cups flour
2 teaspoons baking powder
½ teaspoon salt
½ cup milk
1 teaspoon vanilla

Preheat oven to 350° F. Melt butter in 8- or 9-inch skillet over medium heat. Stir in brown sugar. When thoroughly blended, stir in berries. Cook, stirring, until mixture bubbles and berries begin to give up juice. Add ½ cup white sugar and mash berries slightly. Cook 5 minutes more, then remove from heat. Cream shortening and remaining 1 cup white sugar until light and fluffy. Beat in eggs, one at a time. Sift together dry ingredients and add alternately with combined milk and vanilla to creamed mixture. Blend well, but do not overbeat. Pour batter over blackberry mixture in skillet. Bake 35 to 40 minutes until center of cake springs back when touched lightly. Cool in pan until lukewarm, then turn out onto large serving plate. Serve topped with cream. Serves 6-8.

STRAWBERRY JAM CAKE

Lighter than its blackberry cousin.

1 cup butter, softened
½ cup sugar
1 cup strawberry jam
1 teaspoon ginger
¼ teaspoon cloves

½ cup black coffee
3 eggs, separated
1 teaspoon baking soda
1 heaping tablespoon sour cream
2½ cups flour

Preheat oven to 350° F. Cream butter and sugar until light and fluffy. Stir together jam, spices, and coffee, and beat into creamed mixture. Beat egg yolks well and beat into mixture. Stir baking soda into sour cream and add alternately along with flour to jam-shortening mixture. Beat egg whites until stiff and fold into batter. Bake cake in two 9-inch greased layer pans, for 45 to 50 minutes. When layers test done, remove from oven and allow to cool in pans 10 minutes. Then invert on wire racks. When cool, ice with vanilla or lemon frosting. Serves 8-10.

SOUR CREAM ELDERBERRY CAKE

Nutty, light, and delicious. Try it for a special treat on Sunday morning.

¼ cup shortening
1 cup sugar, white or light brown or
    half and half
2 eggs
2 cups flour
1 teaspoon baking powder
1 teaspoon soda

½ teaspoon salt
1 teaspoon almond extract
1 cup sour cream
2 cups fresh elderberries, or two 10-
    ounce packages frozen, thawed
½ cup chopped pecans, walnuts, or
    almonds

Preheat oven to 350° F. Cream shortening and sugar until light and fluffy. Beat in eggs, one at a time. Sift dry ingredients and add to creamed mixture alternately with combined almond flavoring and sour cream. Grease generously a 10-inch tube pan. Pour half of batter into pan. Sprinkle half of berries on top. Pour in second half of batter, and sprinkle second half of berries on top of that. Sprinkle nuts over all. Bake 55 to 60 minutes, until cake tests done. Remove from oven and cool in pan 10 minutes. Then invert onto serving plate and stripe with glaze while still warm.

Glaze:

1 cup confectioners' sugar
1½ tablespoons water

½ teaspoon almond extract

Mix together and pour over cake.

BLUE ANGEL PUFF

A cake lighter than air.

½ cup shortening
1 cup sugar
2 eggs, separated
½ cup flour
1 teaspoon baking powder

¼ teaspoon salt
⅓ cup milk
1 teaspoon vanilla
1½ cups blueberries, lightly floured

Preheat oven to 350° F. Cream shortening and sugar until light and fluffy. Beat in egg yolks. Sift dry ingredients and add alternately with combined milk and vanilla to creamed mixture. Beat egg whites until stiff and fold in. Fold in blueberries. Pour into greased and floured 10-inch tube pan. Bake 35 to 40 minutes or until cake springs back when touched lightly. Serves 8-10.

TRIFLE OR RASPBERRY RUM CAKE

Digging into this rich confection is no trifling matter. Irresistible.

Fresh or day-old yellow, sponge, or
　layer cake
2 tablespoons rum or sherry
2 cups raspberries, fresh or frozen,
　thawed

½ cup sugar, if berries are
　unsweetened
¼ cup blanched slivered almonds
1 recipe rich egg custard or vanilla
　pudding
1 cup whipping cream

Cut cake into squares or rounds and use them to line bottom and sides of deep dish or mold. Sprinkle cake with rum or sherry. Spoon fruit over cake, sprinkle with almonds, and pour in custard or pudding. Whip cream until stiff and spoon on top of custard. Chill 3 to 4 hours so flavors blend. Serves 6-8.

LAGGTARTA OR SWEDISH PAN CAKE

Lots of fun to make, and the cake itself is the reward.

3 eggs
1 cup sugar
1 cup sifted flour
1 teaspoon baking powder
½ cup butter

1 cup thick applesauce
½ cup pureed raspberries
Confectioners' sugar
2 cups sweetened fresh raspberries

Preheat oven to 425° F. Beat eggs and sugar together until thick and lemon-colored. Sift flour and baking powder and fold into egg mixture. Melt butter and pour *only* clear liquid part slowly into egg and flour mixture, folding it in. Discard white solids of butter. Grease and heat 8-inch skillet. Pour about ⅔ cup batter into skillet, bake 5 to 7 minutes until lightly browned. Gently remove cake and tip out onto rack. Regrease and reheat skillet each time and make 5 more cakes. Combine applesauce and raspberry puree and spread between layers as filling. Sprinkle top of cake with confectioners' sugar. Pour fresh raspberries over all. Serve with sweetened whipped cream. Serves 8-10.

FROSTINGS AND FILLINGS

BLACKBERRY, BLACK RASPBERRY, OR RED RASPBERRY FILLING

Spread between layers of white or spice cake, or fill rich shortcake biscuits. Also makes a pretty and delicious torte mounded between crepes stacked 5 or 6 high.

1 cup whipping cream
½ cup confectioners' sugar
1 cup fresh berries, or one 10-ounce
    package frozen, thawed

2 egg whites
1 tablespoon rum or sherry for black
    berries, or 1 tablespoon kirsch
    for red

Whip cream until stiff, then slowly beat in confectioners' sugar. Puree berries and force through sieve to remove seeds. Beat egg whites until stiff and fold them into whipped cream. Add flavoring to pureed berries, and fold in to cream mixture. Makes about 3 cups.

STRAWBERRY GLAZE

For a strawberry pie or a bright topping for a cheesecake.

3 cups strawberries
⅓ cup sugar
1 tablespoon lemon juice

1 tablespoon cornstarch
(A drop or two of red food coloring)

Hull and crush strawberries, then force them through a ricer or sieve. Pour into saucepan and stir in remaining ingredients. Cook, stirring, over low heat, until thick and clear. Cool thoroughly and spoon over fruit to be glazed.

GLAZE FOR FRUIT PIE, TART, OR COFFEE CAKE

Adds a sweet sparkling fillip that distinguishes any pie, tart, or coffee cake.

3 cups pureed raspberries
1 cup sugar

1 cup light corn syrup

Strain raspberries to remove seeds. Combine puree with sugar and corn syrup in saucepan and cook, stirring, until sugar is dissolved. While glaze is warm, spoon carefully and evenly over cool pastry. Makes enough for several pastries, and stores well in refrigerator.

FRENCH STRAWBERRY FILLING

Spread this between layers of white, yellow, or chocolate cake, and spoon more on top. Superb.

1 cup heavy cream
¼ cup confectioners' sugar
1 egg white

⅛ teaspoon salt
½ teaspoon vanilla
½ cup mashed strawberries

Whip cream until stiff, then slowly beat in confectioners' sugar. Beat egg white until stiff and fold into whipped cream mixture, along with salt and vanilla. Gently fold in mashed strawberries. Makes about 2½ cups.

BERRY JAM FILLING

Spread any jelly or jam made from berries thickly between layers of cake. Then top the cake with mounds of whipped cream or filling made from preceding recipe.

DESSERTS

BAKED BLACKBERRY SANDWICHES

Filled to bursting with juicy good taste, and a sweet bonus of coconut on the outside.

1 4-ounce can sweetened
    condensed milk
12 slices white or brown bread,
    crusts trimmed
1 can flaked coconut (½ cup)

3 cups fresh blackberries
½ cup water
⅓ cup sugar
1 teaspoon vanilla

Preheat oven to 350° F. Brush condensed milk onto both sides of bread slices. Put coconut in bowl, dip bread slices into it, and place 6 slices on greased baking sheet. Cover slices with 2 cups blackberries. Top with reserved slices. Bake 15 to 20 minutes, until lightly browned. Meanwhile, combine remaining cup blackberries with water and sugar in saucepan. Cook, stirring, over moderate heat, until sauce thickens slightly. Remove from heat, stir in vanilla. Place baked sandwiches on serving plates, spoon sauce over them, and serve topped with whipped cream. Serves 6.

BLACK RASPBERRY DELIGHT

Hidden in this dessert is a black raspberry treat.

¼ cup butter
1 cup confectioners' sugar
2 egg yolks
2 egg whites
Pinch of salt
1 package of vanilla cookies, rolled
    into crumbs

2 cups fresh black raspberries, or one
    10-ounce package frozen,
    thawed
¼ cup sugar, if berries not
    pre-sweetened
1½ cups whipping cream

Cream together butter and confectioners' sugar until light and fluffy. Beat in yolks, one at a time. In separate bowl, beat egg whites and salt until stiff. Fold whites into creamed mixture. In greased 9x12-inch pan, spread half of crumbs. Spread cream mixture over them. Top with black raspberries, mashed and sweetened with ¼ cup sugar. Whip cream, spread over berries, and top everything with reserved crumbs. Chill 4 hours or longer before serving. Serves 6-8.

STRAWBERRY FONDUE

Elegant, easy, and lots of fun besides. Try it at a dinner party.

1 quart large ripe strawberries	Creamy Kirsch Sauce
½ frozen pound cake, thawed and cut into cubes	Butterscotch Sauce
	Chocolate Sauce

Heat one of the three sauces (recipes below) in a fondue pot. Dip strawberries and cake into warm sauce. Serves 4.

Creamy Kirsch Sauce:

¾ cup whipping cream	1 or 2 tablespoons kirsch
¾ cup powdered sugar	

Combine cream and sugar in small saucepan and stir until mixture boils. Cook 1 minute. Remove from heat and stir in kirsch.

Butterscotch Sauce:

¾ cup whipping cream	1 or 2 teaspoons vanilla
⅓ cup firmly packed brown sugar	

Combine cream and sugar in small saucepan. Stir until mixture boils and cook, stirring, 1 minute. Remove from heat and stir in vanilla.

Chocolate Sauce:

3 squares semisweet chocolate	¾ cup half-and-half
⅓ cup sugar	1 teaspoon vanilla

Melt chocolate in heavy pan over low heat. Stir in sugar and half-and-half. Cook, stirring, until mixture becomes thick and smooth. Remove from heat and add vanilla.

PAVLOVA

An airy dessert named for the famous Russian ballerina.

4 egg whites	2 teaspoons cornstarch
½ teaspoon salt	3 to 4 cups crushed blackberries
1 cup granulated sugar	1 cup cream, whipped
2 teaspoons vinegar	

Preheat oven to 300° F. Grease an 8-inch cake pan, cover bottom of pan with waxed paper, grease and dust paper with cornstarch. Pour off excess cornstarch. Beat egg whites in large bowl until foamy. Add salt and continue beating until stiff. Add sugar gradually, beating well after each addition. In separate bowl, combine vinegar and cornstarch and blend well. Fold into meringue. Spoon meringue into cake pan, smoothing it to edges of pan and forming an indentation toward center. Place in oven and turn heat down to 250° F. Bake 1 hour and 15 minutes. Remove and let stand on rack until cool, then carefully remove from pan. Spoon crushed blackberries into center of meringue, top with whipped cream, and serve. Serves 8-10.

RED RASPBERRY CONES

Make these with any kind of berry, but be sure to mold the cones when they are still warm. Otherwise, they become brittle and break.

Cones:

2 eggs	1 tablespoon cold water
½ cup sugar	⅔ cup flour

Preheat oven to 350° F. Combine eggs and sugar and beat until thick and lemon-colored. Add water and flour and continue beating until smooth. Grease and flour a baking sheet. Roll dough to ⅛-inch thickness, cut circles out of the dough, 4 inches in diameter. Bake on sheet about 12 minutes, or until edges turn lightly brown. Remove circles one at a time, letting rest stay in oven. Form into cones gently with hands, working quickly so rounds stay pliable. Place cones in airtight container in freezer until ready to serve.

Mousse Filling:

2 egg whites	1 tablespoon lemon juice
Pinch of salt	½ teaspoon vanilla
6 tablespoons sugar	1 cup whipping cream
2 cups red raspberries	

Combine egg whites and salt and beat until stiff. Gradually beat in ¼ cup sugar and beat until stiff and glossy. Crush raspberries and stir in 2 tablespoons sugar, the lemon juice, and vanilla. Whip cream until stiff and fold into meringue; then fold in berries. Place in freezer until firm. Let soften at room temperature before serving. To serve, spoon mousse into cones and top with a fresh berry or two. Makes 6-8 cones.

RASPBERRIES JUBILEE

This makes an elegant dessert, and is not nearly as hard to make as it seems.

¼ cup sugar	½ tablespoon grated lemon rind
1 tablespoon butter	1 cup raspberries
⅓ cup orange juice	1½ tablespoons brandy
¼ cup raspberries, mashed	1½ tablespoons curaçao or Cointreau
1½ tablespoons port wine	liqueur
1½ tablespoons grated orange rind	4 scoops vanilla ice cream
1 tablespoon currant jelly	

In enameled or stainless steel skillet, caramelize sugar and butter over low heat. Stir in orange juice and cook until well mixed. Add ¼ cup berries, mash thoroughly with back of spoon, and stir until well mixed. Stir in port, orange rind, jelly, and lemon rind, and cook until well mixed. Add whole berries and cook, stirring, for 5 minutes. Remove from heat. Add brandy and liqueur, light with a match, and shake pan until flames die. Spoon sauce over ice cream and serve. Serves 4.

FORGOTTEN PUDDING

A melt-in-the-mouth purple and white delight.

5 egg whites
1½ cups sugar
½ teaspoon cream of tartar
½ teaspoon salt

1 teaspoon vanilla
1 cup whipping cream
1 quart black raspberries, crushed
2 teaspoons vanilla

Preheat oven to 450° F. Beat egg whites until stiff. Gradually beat in sugar, beating well after each addition. When egg whites are stiff and glossy, beat in cream of tartar, salt, and 1 teaspoon vanilla. Grease and sprinkle with cornstarch a 9x9-inch pan. Spread meringue in it and set in oven. Immediately turn oven off and let meringue cook, untended, 4 hours. Remove. Whip cream until stiff and spread on meringue. Refrigerate 4 hours or more. Serve cut into squares and topped with crushed fruit mixed with vanilla. Serves 9.

BERRY FOOL

Easy, quick, and very rich. Use any berry for a delicious dessert.

4 cups berries
6 tablespoons shortening, cut in
   small pieces

1 cup sugar
2 cups whipping cream

Heat berries and shortening in saucepan until shortening is melted. Cover and cook slowly for about 10 minutes, until berries are soft. Stir in sugar and cool mixture for 3 to 4 hours. Whip cream until stiff and fold in berries. Chill thoroughly before serving. Serves 6.

WINED BLUEBERRIES

Very easy and extraordinarily good.

1 quart blueberries
1 cup sugar
¼ teaspoon cinnamon or nutmeg
1 cup claret or rosé wine

1 tablespoon honey
2 tablespoons brandy
Sponge cake

Preheat oven to 300° F. Put blueberries in shallow baking dish and sprinkle them with sugar and cinnamon or nutmeg. Bake 15 to 20 minutes, until sugar melts and berries become juicy. Drain off liquid into small saucepan. Pour wine into dish with berries and bake 15 minutes longer. Heat blueberry juice in saucepan with honey and brandy. Boil about 10 minutes, until sauce is thick and syrupy. Pour over berries and serve hot over warm sponge cake. Serves 6.

STRAWBERRIES WITH PEPPER

Sounds terrible, but the pepper adds a tang that perks up the flavor of strawberries.

2 cups strawberries
2 tablespoons sugar
1 tablespoon heavy cream
1 tablespoon brandy

1 tablespoon orange-flavored liqueur
    (curaçao or Grand Marnier)
8 grinds or ¼ teaspoon pepper
¼ cup cream, whipped
2 large scoops ice cream

Crush berries, and into them stir sugar, cream, brandy, liqueur, and pepper. Fold in whipped cream. Spoon over ice cream and serve. Serves 2.

STRAWBERRY FLUFF

Light, sweet, and airy.

2 cups fresh strawberries
2 tablespoons confectioners' sugar
½ pound small marshmallows, or 18
    large marshmallows, cut up

1 cup whipping cream
½ teaspoon almond extract
½ teaspoon vanilla extract

Slice berries and sprinkle with sugar. Stir in marshmallows and let sit 1 to 2 hours, until thick. Whip cream and flavorings and fold into berries. Chill thoroughly and serve. Serves 6.

CREPES WITH BLUEBERRY-LICHEE FILLING

Lichee nuts give an exotic touch to this attractive combination.

Crepes:

¾ cup flour, sifted before measuring
½ teaspoon salt
1 teaspoon baking powder
2 tablespoons confectioners' sugar
2 eggs

⅔ cup milk
⅓ cup water
½ teaspoon vanilla or ½ teaspoon
  grated lemon rind

Sift together dry ingredients. In a separate bowl, beat eggs, then beat in milk, water, and flavoring. Make a well in the dry ingredients and pour in the liquid. Stir together with a few strokes. Batter will be lumpy. Grease lightly a 5-inch skillet. Heat it, and spoon in a small amount of batter. Tip pan so batter coats bottom. When batter is light brown underneath, turn it and brown other side. Stack crepes, covered, until ready to use. Makes about 12 crepes.

Filling:

2 cups blueberries, mashed
8 large canned lichees, sliced thin
½ cup sugar

1 tablespoon dark rum
½ cup confectioners' sugar
1 cup cream, whipped

Mix together blueberries, lichees, granulated sugar, and rum. Let sit 1 to 2 hours to meld flavors. Spoon about 3 tablespoons filling onto each crepe and roll crepe up over filling. Sprinkle with confectioners' sugar, top with whipped cream, and serve. Serves 4-6.

CREAMY ELDERBERRY DESSERT

Usually this dish is made with strawberries. But elderberries add a whole new dimension.

1 cup elderberries
18 large marshmallows, cut in
  quarters

1 cup whipping cream
1 tablespoon rum

Combine and stir elderberries and marshmallows in bowl. Let stand 1 to 2 hours, until thickened. Whip cream. Stir rum into berries, then fold in cream. Serve thoroughly chilled. Serves 6-8.

RASPBERRY CHARLOTTE

Feather-light, and a gorgeous shade of pink besides.

2 packages (3 ounces each) raspberry
    gelatin mix
1 cup cream, whipped
1 cup raspberry jam
1 cup cream, whipped
2 cups fresh raspberries, or one
    10-ounce package frozen, thawed

Prepare gelatin according to package directions and chill until slightly thickened but not set. Beat with electric beater until thick and fluffy. Whip cream and fold in. Fold in jam. Rinse large mold in cold water and pour mixture into it. Chill until firm. Unmold, top with second cup whipped cream and raspberries. Serves 16.

PORTED BLUEBERRY CRISP

Blueberries never ever tasted so good. Use either fresh or frozen berries with excellent results.

1 quart blueberries
¼ cup port wine
¾ cup sifted flour
¾ cup firmly packed brown sugar
½ cup shortening, cut into bits
1 teaspoon nutmeg

Preheat oven to 375° F. Combine blueberries and port in greased shallow baking dish. In separate bowl, stir remaining ingredients with a fork until mixture resembles coarse meal. Sprinkle over berries. Bake 30 to 40 minutes, until golden. Serve with cream or ice cream. Serves 6-8.

FROZEN LEMON SOUFFLE WITH BLUEBERRIES

The tart lemon soufflé complements the sweet taste of blueberries in this airy combination.

2 cups boiling water
2 packages (3 ounces each) lemon
    gelatin mix
2 cups lemonade
¼ cup lemon juice
2 teaspoons grated lemon rind
1 pint cream, whipped
2 cups blueberries, crushed
½ cup sugar
1 teaspoon vanilla

Pour boiling water over gelatin and stir until gelatin dissolves. Add lemonade, lemon juice, and lemon rind. Chill until slightly thickened. Beat until fluffy and thick. Fold in whipped cream. Rinse out 1½-quart soufflé dish and pour mixture in. Fill to within 1 inch of top, place a 3-inch collar of waxed paper around inside of dish, and pour in rest of soufflé. Chill until firm. Remove collar. Stir sugar and vanilla into berries and use as sauce for soufflé. Serves 8.

BLACKBERRY DUMPLINGS

Rich and delicious. Make it with frozen berries for a hearty midwinter treat.

1 quart blackberries	2 teaspoons baking powder
1½ cups water	¼ teaspoon salt
1 cup sugar	3 tablespoons shortening
½ teaspoon salt	⅓ cup milk
1 cup flour	

In saucepan, combine blackberries, water, sugar, and salt. Bring to boil and simmer 10 minutes. In bowl, combine dry ingredients and cut in shortening until mixture resembles coarse meal. Stir in milk until barely blended. Drop dumpling mix by spoonfuls into simmering berries. Cover tightly and cook on top of stove for 10 minutes. Serves 4-6.

ELDERBERRY GRUNT

An interesting twist on an old favorite.

2 cups elderberries, stems removed	½ cup sugar
½ cup water	1 package refrigerated biscuit dough,
1 tablespoon lemon juice	or 1 recipe biscuit dough
¼ teaspoon allspice	

Combine elderberries, water, lemon juice, allspice, and sugar. Stir until berries are soft and sugar is dissolved. Put into greased deep baking dish. Cover with biscuit dough. Set dish in pan of boiling water on top of stove. Water should be at level of mixture in bowl. Cover pan and cook on top of stove for 1 hour. Check water level occasionally and add more if necessary. Serve topped with whipped cream. Serves 6-8.

BERRY CRISP

A terrific way to spice up fresh or frozen berries.

1 quart berries	½ teaspoon salt
1 tablespoon cinnamon	1 teaspoon baking powder
1 cup brown sugar	1 egg
1 cup white sugar	2 tablespoons butter
1 cup flour	

Preheat oven to 350° F. In greased baking dish, combine berries, cinnamon, and brown sugar. Stir to blend. In separate bowl, combine dry ingredients and stir in egg until mixture becomes crumbly. Sprinkle over berries. Dot top with butter. Bake 40 to 50 minutes, until mixture is bubbly and browned. Serve with ice cream. Serves 4-6.

LATE SUMMER PUDDING

This has all the complexity of a Shakespearean comedy. And is equally delightful.

2 cups blackberries
½ cup sugar
8 slices bread, white or brown, crusts trimmed
1 tablespoon flour
6 tablespoons sugar
½ teaspoon salt
3 egg yolks
1½ cups milk
1 teaspoon vanilla
½ cup whipping cream

Combine blackberries and sugar in saucepan and cook, stirring, over low heat until soft. Use bread slices to line a 1-quart mold, saving enough bread to cover top as well. Pour fruit and juice into mold and top with reserved bread. Place saucer on top and a heavy can on saucer. (This forces bread to absorb juice.) Store in refrigerator for 8 hours or longer. Meanwhile, combine flour, sugar, and salt in saucepan. Beat egg yolks slightly and add, with milk, to flour mixture. Cook over medium heat, stirring constantly, until slightly thick and smooth. Or cook in double boiler over hot water, stirring, until thick and smooth. Do not let custard boil. Cool slightly and stir in vanilla. Chill thoroughly. At serving time, whip cream and fold gently into custard. Unmold pudding onto serving plate, spoon a little custard over it, and serve remaining custard in sauceboat on the side. Serves 4-6.

MAINE BLUEBERRY PUDDING

Traditionally made with Maine's famous lowbush semi-wild berries, this can also be made with highbush or rabbiteye varieties.

3 cups blueberries
1 teaspoon cinnamon
¾ cup sugar
½ cup water
6 slices bread, crusts trimmed

Combine berries, cinnamon, sugar, and water and cook over moderate heat 10 minutes. Layer bread and berries in 4x8-inch loaf pan. Chill several hours. Serve in slices topped with cream. Serves 6-8.

RASPBERRY PUDDING

An Old World treat that has fortunately migrated to the New World.

2 cups fresh raspberries, or two 10-ounce packages frozen, thawed
1 cup sour cream
2 eggs
1 tablespoon sugar
1 tablespoon flour

Preheat oven to 350° F. Place raspberries in a deep, greased 1-quart baking dish. Put in oven and bake 10 minutes, until berries release juice. Beat remaining ingredients together until well blended. Pour over berries. Bake 40 to 45 minutes, until firm and lightly browned. Serve either warm or cooled to room temperature. Serves 4.

BAVARIAN ELDERBERRY CREAM

Use a decorative mold, top with Red Raspberry Sauce, and create a colorful sensation.

1 quart elderberries
1 cup sugar
1 tablespoon gelatin, unflavored

3 tablespoons water, *and* 3 tablespoons boiling water
1 tablespoon lemon juice
1 cup whipping cream

Crush the elderberries and stir in the sugar. Let stand 30 minutes. Soak gelatin in cold water 10 minutes, then add boiling water to dissolve gelatin. Stir into berries, along with lemon juice. Cool until thick but not quite set. Whip cream and fold gently into berries. Pour into mold and chill 12 hours or longer to insure its firmness. Unmold and serve with Red Raspberry Sauce (see Sauces, p. 115) and ginger cookies. Serves 8-10.

BLACK RASPBERRY SCORCHED CREAM

Crunchy hot meringue conceals cool black raspberry custard.

4 cups black raspberries
2 teaspoons vanilla
5 eggs, separated

½ cup brown sugar
2 cups milk, preferably whole milk
¼ cup white sugar

Put berries in greased 1½-quart baking dish and sprinkle vanilla over them. In top of double boiler, beat yolks until light. Beat in brown sugar. Scald milk, stir a little of yolk mixture into milk, then pour milk into yolk mixture, beating constantly. Cook in double boiler, stirring constantly, over hot water, until custard is thick and smooth. Cool 1 to 2 hours. Pour over black raspberries, cover, and chill thoroughly. At serving time, beat egg whites until stiff, then gradually beat in 2 tablespoons sugar. Spread meringue over berries and custard. Sprinkle top with remaining 2 tablespoons sugar. Brown, *carefully*, under broiler, keeping top about 3 inches from flame. Do not burn. Serve while meringue is hot. Serves 6-8.

RASPBERRIES ROMANOFF

Simple, quick, and absolutely irresistible. Try using coffee ice cream for a tasty change.

2 cups vanilla or chocolate ice cream, softened
1 cup cream, whipped
1 tablespoon lemon juice

¼ cup cream sherry
2 quarts raspberries
½ cup confectioners' sugar

Soften ice cream by beating with a fork. Fold whipped cream into it, then the lemon juice and sherry. Crush berries, stir in confectioners' sugar, and fold into cream mixture. Serve immediately. Serves 8.

RASPBERRY SPIRAL MOLD

Takes time but is well worth the effort.

1 cake for jelly roll, cut in half
   crosswise
1 cup raspberry jelly
2 tablespoons kirsch
3 cups fresh raspberries, or two
   10-ounce packages frozen,
   thawed and drained, reserving
   juice

4½ teaspoons unflavored gelatin
⅓ cup kirsch
5 egg yolks
¾ cup granulated sugar
1½ cups milk, scalded
2 teaspoons vanilla
3 egg whites
1 cup whipping cream

Spread each half of cake with jelly that has been melted and combined with 2 tablespoons kirsch. Beginning with a long side, roll up each half of cake and wrap each roll tightly in foil. Chill 1 hour or longer. Force berries through sieve and reserve puree. In small bowl, sprinkle gelatin over kirsch to soften. In large bowl, beat together egg yolks and sugar until mixture is thick. Pour in scalded milk in constant stream, while beating, and transfer mixture to heavy saucepan. Cook, stirring, over low heat until mixture is slightly thickened; mixture will coat spoon lightly. Do not simmer or boil. Remove from heat and stir in gelatin mixture and vanilla. Stir until gelatin dissolves completely. In another bowl, beat egg whites until stiff and fold into custard. Set custard in bowl of ice and stir occasionally until cool and thickened. Remove from ice.

To assemble, cut cake rolls into ⅓-inch slices. Whip cream until stiff and fold reserved raspberry puree into it. Rinse 2-quart mold with cold water and oil lightly. Line bottom with waxed paper, cut to fit. Cover bottom of mold with slices of jelly roll. Now fold raspberry-cream mixture gently into custard. Pour 1 inch of this over jelly roll slices, and continue to layer cake and custard until level in mold is 2½ inches from top. Stand remaining slices of jelly roll against sides of mold and pour in remaining custard mixture. Cover tightly and chill 4 hours or more, until custard is thoroughly set.

To unmold, run a knife carefully around edge of mold, dip mold in hot water for a few seconds, put serving plate face down over mold, and invert mold. Rap bottom sharply to dislodge mold. Remove waxed paper and serve spiral mold with Red Raspberry Sauce (see Sauces, p. 115).

RASPBERRY POSSETT

Another traditional English dish, this dessert puts leftover bread to excellent use.

4 cups red raspberries
1 cup sugar
1 teaspoon vanilla

4 slices bread, buttered on both
   sides and cut into cubes

Simmer berries and sugar over medium heat for 10 minutes, stirring, until berries are soft and juicy. Remove from heat and stir in vanilla. Place bread cubes in bowl and pour berry mixture over them. Stir gently to mix. Let stand several hours so bread becomes saturated with juice. Serve with cream or whipped cream. Serves 6-8.

BLACKBERRY RICE PUDDING

A terrific way to turn leftover rice into a real delicacy. Make this either at the height of blackberry season, or in deep winter with frozen berries.

1 quart blackberries, or three
   10-ounce packages frozen,
   thawed
½ cup sugar, white or brown

2 tablespoons dark rum or 1
   teaspoon vanilla
3 cups cooked rice
2 cups cream, whipped

Crush berries and stir in sugar and vanilla. Let stand 10 to 15 minutes. Stir in rice. Whip cream and fold gently into berries. Pour into serving dish, chill several hours, and serve. Serves 8-10.

FINNISH ELDERBERRY SNOW

Rich in the deep purple taste of elderberries.

2 cups elderberries, stemmed
¾ cup sugar
1 tablespoon grated orange rind
4 egg whites

Pinch of salt
¾ cup cream, whipped
Spice cake

Crush elderberries and stir in sugar and orange rind. Beat egg whites and salt until stiff. Whip cream. First fold berries, then whipped cream into egg whites. Mound snow in serving bowl and refrigerate, if necessary, until serving. Serve with wedges of spice cake. Serves 6-8.

RED PUDDING

Traditional Scandinavian fare.

2 cups raspberries, crushed
3 tablespoons sugar
1 tablespoon cornstarch

¼ cup berry juice
½ cup toasted slivered almonds

Simmer berries over moderate heat for 5 minutes until very soft. Mash fruit and force through sieve into another saucepan. Add sugar. Mix cornstarch with berry juice and stir into puree. Cook, stirring, over moderate heat until pudding thickens. Spoon into four dishes, sprinkle with almonds, and chill. Serve topped with sweetened whipped cream. Serves 4.

BERRY COMPOTE

One of any number of possible combinations that makes use of all berries in season. Tastes and colors complement each other.

1 cup each of 3 or more fresh berries
1 cup cantaloupe or other melon
    slices

Sugar to taste — ½ cup or more

Stir all ingredients together and let stand 1 or 2 hours. Serve with ginger cookies. Serves 6-8.

SUGAR-NUT BERRY DESSERT

Not quite a pie and not quite a bar, this defies any description except "scrumptious."

Crust: ½ cup softened butter
1½ cups flour
⅓ cup powdered sugar

Preheat oven to 350° F. Mix flour, powdered sugar, and butter until mixture resembles coarse meal. Press into bottom of 9x13-inch pan. Bake 15 minutes and let cool on rack.

Filling:
1 quart fresh blackberries
2 cups sugar   ½ teaspoon baking powder
2 eggs   1 teaspoon vanilla
½ teaspoon salt   ¾ cup chopped pecans or walnuts
¼ cup flour   1 cup whipping cream

Crush berries slightly. Stir in 1 cup sugar and let sit to form juice. Drain, reserving juice. Beat eggs with remaining 1 cup sugar until light and fluffy. Beat in salt, flour, baking powder, and vanilla, and beat well. Spread crust with drained berries. Sprinkle berries with chopped nuts. Pour egg mixture over all, and bake 30 minutes, or until golden. Cool.

Sauce:
½ cup reserved blackberry juice
½ cup water   2 tablespoons cornstarch
½ cup sugar   1 tablespoon lemon juice

Combine ingredients in saucepan and cook, stirring, until thick and clear.

To serve, cut pastry into squares. Whip cream and spoon it over squares. Top with sauce and sprinkle with nuts. Serves 10.

BERRIES SUPREME

Spooned generously over rich cake or ice cream, this makes a simple but irresistible finish to any meal.

1 quart ripe berries   3 to 5 tablespoons superfine
1 teaspoon vanilla or almond extract   granulated sugar

Place 1 cup berries in blender and puree. Gradually blend in flavoring and sugar to taste. Blend only until puree is consistency of thick cream. Pour puree over remaining berries. Mix gently. Chill several hours before serving. Serves 4.

STEAMED BLUEBERRY PUDDING

Terrific staver-off of midwinter chills and melancholy. Equally good with fresh fruit in high summer.

½ cup butter
½ cup sugar, white or light brown
1 egg
1½ cups flour
2 teaspoons baking powder

½ teaspoon salt
⅓ cup milk
Grated rind of ½ lemon
1 cup blueberries, fresh or frozen

Cream butter and sugar until light and fluffy. Beat in egg. In separate bowl, combine dry ingredients and add to creamed mixture alternately with milk. Stir in lemon rind. Fold in blueberries. Grease 1-quart mold and pour batter into it. Place mold on rack in large saucepan, and pour boiling water into saucepan so level reaches halfway up side of mold. Cover saucepan tightly and steam 1 hour. Unmold pudding and serve hot with lemon or hard sauce. Serves 6.

BLUEBERRY TANSY

A traditional and old-fashioned way to do up fresh blueberries.

½ cup butter
2 cups blueberries
2 egg yolks, beaten
½ cup whipping cream

½ cup sugar, white or brown
2 tablespoons lemon juice
½ teaspoon nutmeg

In skillet, melt butter and cook berries in it until they become soft. In bowl, combine beaten egg yolks and lightly whipped cream. Stir gently into simmering berries. Stir in sugar and continue to cook, stirring, until tansy is slightly thick. Spoon into serving dish, sprinkle with lemon juice and nutmeg. Serve warm or cold. Serves 4-6.

MULLED ELDERBERRY PUDDING

A spicy taste with deep color to warm you up on a winter night.

2 cups frozen elderberries, thawed
1 cup water
1 cup grape juice
¾ cup sugar

1 stick cinnamon
½ teaspoon allspice
2 tablespoons cornstarch
¼ cup water

Combine elderberries with water and cook over low heat for 10 minutes. Do not boil. Add grape juice, sugar, and spices and boil slowly for 15 minutes. Mix cornstarch in water and stir slowly into juice mixture. Cook, stirring, until pudding becomes thick and clear. Remove from stove and pour into serving dishes. Cool. Serve with whipped cream. Serves 6.

FLUMMERY

A silly name for a superb dessert.

2 cups blackberries
½ cup sugar
2 tablespoons lemon juice
1 tablespoon cornstarch

¼ cup water
2 tablespoons cream sherry
Slices of pound cake or ladyfingers
Whipped cream

Combine berries, sugar, lemon juice, and cornstarch mixed with water in saucepan. Cook, stirring, over moderate heat until mixture boils and thickens. Remove from heat and stir in sherry. Pour hot sauce over sliced cake or split ladyfingers. Serve with whipped cream. Serves 4-6.

FILLED MARZIPAN COOKIES

Make these around Christmas for a real European delicacy.

½ cup butter
¼ cup sugar
1 egg yolk
3 tablespoons and ½ cup almond
    paste, separated

1½ cups sifted cake flour
1 egg white
1 cup strawberry jam

Preheat oven to 350° F. Cream butter, then beat in sugar, yolk, and 3 tablespoons almond paste. Add flour, blend well, wrap tightly in aluminum foil or plastic wrap, and chill. Work egg white into ½ cup almond paste and set aside. Roll chilled dough out to ¼-inch thickness. Cut into shapes and place 1 inch apart on greased baking sheets. Make a circle on each shape with reserved paste-egg white mixture by forcing mixture through pastry tube. Fill each circle with jam or jelly. Bake 15 minutes. Makes 16-20 good-sized cookies.

SYLLABUB

Syllabub, sometimes served as a milk punch instead of in this rich version, dates from early colonial days in this country. In England, it was popular long before that.

1 cup heavy cream
½ cup confectioners' sugar
2 egg whites
2 tablespoons cream sherry

2 cups fresh raspberries or
    blackberries
½ cup slivered toasted almonds
White cake

Whip cream until stiff, and then gradually beat in ¼ cup confectioners' sugar. Beat egg whites until foamy, then add remaining ¼ cup sugar very slowly while beating whites stiff. Fold together the two mixtures. Fold in sherry and pour over berries. Sprinkle top with almonds. Serve with white cake. Serves 4-6.

COEURS A LA CREME WITH RASPBERRY SAUCE AND CREME FRAICHE

Divine on Valentine's Day (or any other day).

5 ounces cream cheese, softened	2 cups fresh raspberries, crushed, or
½ teaspoon vanilla extract	two 10-ounce packages frozen,
¼ cup confectioners' sugar	thawed and drained
1 cup heavy cream	¼ cup unsweetened apple juice
	¾ cup currant jelly

Beat cream cheese until light. Beat in vanilla, and beat in sugar gradually. Whip cream until stiff and fold into cheese mixture. Rinse 6 pieces of cheesecloth in cold water and use them to line 6 small heart-shaped molds. Spoon the cheese mixture into the molds. Overlap ends of cheesecloth and draw up tightly over each mold. Chill thoroughly. Meanwhile, drain raspberries. Combine apple juice and jelly and cook, stirring, over low heat until smooth. Pour over raspberries. Turn out cheese molds onto chilled plates and top with raspberry sauce and crème fraîche. Serves 6.

Crème Fraîche:

1 cup heavy cream	1 tablespoon buttermilk

Combine cream and buttermilk and place in tightly sealed jar. Shake and let stand at room temperature 8 to 10 hours, until cream has thickened.

GAUFRES DE LIEGE OR BELGIAN WAFFLES

Not at all the waffles Americans eat for breakfast. Rather, a hearty snack or dessert to be long remembered.

1 package active dry yeast	7 tablespoons shortening, melted
2½ cups lukewarm milk	and cooled
3 cups flour	2 egg whites
¼ cup sugar	Confectioners' sugar
⅛ teaspoon salt	1 cup whipping cream
2 egg yolks	2 cups fresh strawberries, hulled and
1 teaspoon vanilla	sliced

Sprinkle yeast over ½ cup warm milk in large mixing bowl. Stir to blend and let sit 5 minutes. Pour in remaining milk and stir in flour, sugar, and salt. Stir until smooth. Stir in egg yolks, vanilla, and shortening. Cover bowl with damp kitchen towel and set to rise in warm, draft-free place for about 30 minutes. Preheat waffle iron. Beat egg whites until stiff. Stir down yeast batter and gently fold in beaten whites. Bake batter, about 1½ cups at a time, for about 5 minutes in waffle iron. Try not to open iron until waffle is done. Serve waffles hot, sprinkled with confectioners' sugar and piled high with strawberries and whipped cream. Serves 4-6.

FROZEN DESSERTS

RASPBERRY OR STRAWBERRY SHERBET

Very easy, and the perfect complement to fresh fruit cup or sugar cookies.

2 teaspoons gelatin
¼ cup cold water
1 quart fresh berries
¼ cup lemon juice

1¾ cups water
¾ cup sugar
2 egg whites

Soak gelatin in cold water. Press berries through sieve or puree in blender. Stir lemon juice into puree. Combine water and sugar and boil for 10 minutes. Add gelatin mixture and stir to dissolve. Cool syrup. Stir in berries. Chill thoroughly. Beat egg whites until stiff but not dry and fold into berry mixture. Pour into flat trays and freeze until slushy. Beat with electric mixer to break up large crystals, return to freezer, and freeze until firm. Serves 4-6.

BLACK RASPBERRY BUTTERMILK ICE CREAM

The taste and texture of this cleverly disguise its caloric economy.

2 cups black raspberries, pureed in
   blender
3 tablespoons lemon juice

⅛ teaspoon salt
1½ cups buttermilk
½ cup sugar

Strain puree through sieve to remove seeds. Combine all ingredients and stir until sugar is dissolved. Pour into tray, freeze until texture is slushy. Put mixture in bowl and beat with electric beater until texture is smooth but not melted. Return to tray, replace in freezer, and freeze until solid. Soften slightly before serving. Serves 4-6.

BLACKBERRY SHERBET

Top this with fresh buttermilk, serve molasses cookies on the side, and the result will astound you.

6 cups blackberries
½ to ¾ cup sugar
1 teaspoon vanilla

2 egg whites
2 tablespoons sugar

Force blackberries through sieve or puree them in blender. Strain to eliminate some seeds. Stir in sugar and vanilla. Freeze in bowl or tray until mushy. Beat with electric beater to break up large ice crystals. Beat together egg whites and 2 tablespoons sugar until stiff. Fold into blackberry mixture and freeze until firm. Let soften slightly before serving. Serves 6-8.

RHUBARB-STRAWBERRY SHERBET

Perfect for making and eating at the height of rhubarb-strawberry season. Also makes up deliciously with frozen fruit in the off-season.

5 cups rhubarb, cut up 2½ cups superfine sugar
1 cup water 2 cups strawberries

Combine rhubarb and water and simmer, covered, in saucepan for 5 minutes or until rhubarb is soft. Puree rhubarb and sugar in blender until all sugar is dissolved. Puree strawberries separately. Combine strawberry puree and rhubarb puree and blend well. Pour into loaf pans or freezing trays and freeze until mushy. Scrape sherbet into bowl and beat with electric mixer until smooth but not melted. Return mixture to pans and continue freezing a second time until mushy. Repeat beating process, return to freezer in covered container, and freeze until solid. Soften slightly before serving. Serves 6-8.

RASPBERRY ICE CREAM

The absolute essence of rich ice cream. Splendid made with fresh or frozen fruit.

2 quarts fresh berries, or six 2¼ cups sugar (use less for frozen
   10-ounce packages frozen,   berries)
   thawed and drained 4 cups (1 quart) heavy cream
1 teaspoon salt

Mash berries and stir in sugar. Let stand 30 minutes, then strain to remove seeds. Measure 3½ cups juice and combine with cream and salt in ice cream freezer. Freeze as directed. Makes about 2 quarts ice cream.

STRAWBERRY BOMBE

Very elegant, and surprisingly simple. Try different combinations of sherbet and ice cream.

1 quart strawberry sherbet 1 cup whipping cream
1 quart strawberry ice cream Whole perfect strawberries

Use strawberry sherbet to line a 2-quart mold. Fill mold with soft ice cream. Cover and freeze until firm. When ready to serve, whip cream and pile on unmolded bombe. Garnish with whole strawberries.

FROSTY BLUEBERRY FLUFF

Light and cool and crunchy. Perfect any time of year, with fresh or frozen berries.

Crumb Crust:

1 cup flour
¼-cup brown sugar
1 teaspoon cinnamon

½ cup shortening
½ cup chopped pecans or walnuts

Combine all ingredients; spread in pan. Bake at 350° F. about 20 minutes, until brown and crisp. Stir every 5 minutes. Mixture will be very crumbly. Cool.

Filling:

2 egg whites
1 cup sugar
2 cups blueberries, fresh or frozen, thawed

2 tablespoons lemon juice
1 cup whipping cream

Beat together all ingredients *except cream* in large bowl for 10 minutes at high speed, until very light and fluffy. Whip cream until stiff, and fold into blueberry mixture.

To assemble, spread ⅔ of crumbs on bottom of 9x14-inch pan. Gently spoon in fluff. Sprinkle with remaining crumbs. Freeze until firm. Serves 6-8.

STRAWBERRY ICE CREAM RICHESSE

The richest of the rich, this is really a melt-in-the-mouth frozen custard.

3 quarts strawberries
1¾ cups sugar
8 egg yolks
¼ teaspoon salt

4 cups milk
1 teaspoon vanilla extract
4 cups heavy cream

Hull and crush berries. Stir in ¼ cup sugar and set aside. Combine egg yolks and 1½ cups sugar in bowl and beat until thick and lemon-colored. Beat in salt. Heat milk until it almost boils. Add milk very gradually to egg yolk mixture, beating constantly. Transfer mixture to saucepan and cook over very low heat until mixture coats spoon lightly. Stir constantly. Cool custard and stir in vanilla. When thoroughly cool, stir in strawberries and cream. Freeze in ice cream freezer according to manufacturer's directions. Serves 18 or more.

BLACK RASPBERRY MOUSSE

Any berry will work when this recipe is used.

1 cup black raspberries
2 egg whites
⅛ teaspoon salt

½ cup sugar
1 cup heavy cream

Puree black raspberries and strain to remove some seeds. Beat egg whites and salt until stiff, then beat sugar into them a few tablespoons at a time. Whip cream until stiff. Fold berry puree into meringue, then fold in cream. Chill in bowl or in individual stemmed glasses at least 3 hours. Serves 6.

CLASSIC RASPBERRY PARFAIT

Substitute other berries and flavorings, such as blackberries with rum.

2 cups sugar
2½ cups water
1 cup raspberries
6 egg yolks

½ teaspoon vanilla extract
2 tablespoons framboise liqueur
1 cup whipping cream
Toasted slivered almonds

Combine sugar and water and bring to a boil. Boil 5 minutes, or until candy thermometer registers 220° F. Stir berries into syrup and simmer 2 minutes, or until berries are slightly softened. Remove berries with slotted spoon and puree them. Measure ¾ cup syrup and set aside. (Store remaining syrup for another use.) In a heavy saucepan over very low heat, combine ¾ cup syrup and egg yolks. Whisk or beat vigorously until mixture becomes light and fluffy. Do not overcook. When mixture is fluffy and very warm, remove from heat and beat with electric mixer until mixture is cool. Gradually beat in pureed berries, vanilla, and framboise. Beat cream until stiff, stir a large spoonful into berry mixture, then fold the rest in gently. Chill in bowl or individual serving dishes in freezer for at least 2 hours. Serve garnished with toasted almonds. Serves 6-8.

ARCTIC RASPBERRY MOUSSE

Halfway between regular mousse and ice cream.

2 cups raspberries
½ cup sugar

½ tablespoon lemon juice
2 cups heavy cream

Crush berries. Stir sugar and lemon juice into them. Whip cream until stiff and fold into berries. Turn gently into 2-quart mold or deep tray. Freeze until almost firm, about 2 hours. Serves 8.

ELDERBERRY ICE CREAM

An unusual and delicious version of ice cream.

⅔ cup sugar
1 tablespoon lemon juice
⅓ cup water
¼ teaspoon cream of tartar

2 egg whites
1½ cups heavy cream
1½ cups elderberries, slightly
    crushed

Combine sugar, lemon juice, and water and bring to a boil in small saucepan. Stir until sugar is entirely dissolved. Stir in cream of tartar. Cover pan and continue to boil 10 minutes, or until syrup spins a thread when dropped into cold water. Beat egg whites until foamy. Pour syrup slowly into egg whites, beating constantly, and continue beating until mixture is cool. Stir in cream and elderberries and freeze in ice cream maker according to manufacturer's directions. Serve at once. Serves 6-8.

SAUCES

BLUEBERRY SAUCE AU RHUM

Marvelous on either plain or blueberry pancakes, on lemon or orange soufflé, or on vanilla custard or ice cream.

3 cups blueberries, fresh or frozen, unsweetened
⅔ cup sugar
½ cup water
¼ cup lemon juice
2 tablespoons dark rum

Combine all ingredients except rum in saucepan and simmer, covered, for 5 minutes. Put mixture into blender and blend until pureed, if desired. Return to saucepan and simmer 7 minutes longer, uncovered. Remove from heat, stir in rum, transfer to serving bowl, and serve hot. Makes about 2½ cups.

RED OR BLACK RASPBERRY SAUCE

Terrific on pancakes, waffles, crepes, plain cake, or ice cream, and especially on any red berry dessert.

2 cups red or black raspberries, fresh or frozen, thawed and drained
⅓ to ½ cup sugar
2 tablespoons rum or 1 teaspoon vanilla

Place ingredients in blender or food processor and puree them. Force puree through a sieve to remove seeds. Sauce may be served cold or hot. Makes about 2 cups.

ORANGE STRAWBERRY SAUCE

Blends well with the flavors of fruit or ice cream. Or try it over berry-filled crepes.

3 cups sliced strawberries
1 cup orange juice
¼ cup sugar
1 tablespoon grated orange peel

Place all ingredients in blender or food processor and blend until pureed. Makes about 2 cups.

ELDERBERRY SAUCE

Dip pieces of Camembert or other cheese in egg-and-flour batter and deep fry them. Then top with this rich purple sauce, which also complements lemony or vanilla-flavored desserts.

⅔ cup elderberries, stemmed and crushed
½ cup water
1 teaspoon vanilla

Dash of cinnamon
Dash of nutmeg
2 tablespoons chopped toasted almonds

Place elderberries and water in small saucepan and stew, covered, for 10 minutes, or until elderberries become soft and juicy. Remove from heat and stir in vanilla and spices. Cool before serving, or serve hot, sprinkled with almonds. Makes about 1¼ cups.

AIRY STRAWBERRY SAUCE

Serve this with fruit cup or plain cake. Looks and tastes like a dream.

1 egg white
⅛ teaspoon salt
1 cup crushed strawberries

1 cup confectioners' sugar
1 tablespoon lemon juice
(2 tablespoons orange-flavored liqueur)

Beat egg white with salt until stiff. Beat in strawberries and continue to beat until mixture is fluffy. Gradually beat in sugar and lemon juice until thick. If desired, fold in liqueur. Serve immediately. Makes about 3 cups.

STRAWBERRY HARD SAUCE

Try this on pancakes or waffles or bubbling-hot blueberry desserts.

½ cup shortening
2 cups confectioners' sugar

1 teaspoon vanilla or 1 tablespoon brandy
1 cup crushed strawberries

Beat shortening until creamy. Beat in confectioners' sugar, a little at a time, until mixture is fluffy. Gradually beat in flavoring. Crush hulled strawberries and fold in gently. Chill, covered, for 1 hour. Makes about 2½ cups.

JAMS AND JELLIES

CHERRY-BLACK RASPBERRY CONSERVE

Sometimes a combination of fruits surpasses either fruit by itself. This is a truly great duo.

3 cups sour pie cherries
⅓ cup water
3 cups black raspberries

4½ cups sugar
¾ cup chopped blanched almonds

Pit cherries and combine with water in saucepan. Cook 5 minutes, or until tender. Stir in black raspberries and sugar and cook until mixture is thick and clear. Stir frequently while cooking. Add almonds and cook 5 minutes longer. Pour into hot sterile jars and seal. Makes about six 8-ounce jars.

A-PLUS RASPBERRY CONSERVE

Better-than-the-average conserve for eating with meats or muffins.

2 cups red raspberries
2 cups strawberries
1 cup raisins
3 cups rhubarb, cut into small pieces

2 oranges, sliced thin
2 lemons, sliced thin
Sugar
2 cups chopped walnuts

Put fruit into kettle and add half as much sugar as there is fruit. Bring slowly to a boil and continue to cook at a simmer, stirring occasionally, until mixture thickens, about 30 minutes. Stir in walnuts, pour into hot sterile jars, and seal. Makes about 2 quarts.

BLUEBERRY BUTTER

Thick and spicy, this turns plain toast into a treat.

2 quarts fresh blueberries
8 large cooking apples, peeled,
   cored, and chopped
8 cups sugar, white or light brown

1 teaspoon ground allspice
1 teaspoon ground mace
1 teaspoon cinnamon
1 teaspoon nutmeg

Combine ingredients in large pot or kettle and stir over moderate heat until mixture boils. Turn heat to low and simmer, uncovered, 1 hour, or until mixture becomes thick. Stir occasionally to prevent sticking. When mixture is thick, spoon into hot sterile jars and seal. Makes about 8 pints.

DOUBLE RED STRAWBERRY JAM

An easy way to create jam laden with plump whole berries.

1 quart whole ripe strawberries   Juice of ½ lemon
4 cups sugar

Place strawberries in deep pot and pour sugar and lemon juice over them. Stir very gently with wooden spoon over low heat until berries become juicy. Turn heat up to medium and stop stirring. When mixture boils all over, cook for exactly 15 minutes without stirring. Remove mixture from heat at end of period, and spoon gently into hot sterile jars. Seal and store a week or more before consuming. Makes 4-5 cups jam.

BERRY CORDIAL JELLY

Richer-tasting than jelly made with unfermented berry juice. Try it with framboise liqueur, strawberry liqueur, blackberry liqueur, or cassis.

2 cups berry liqueur   ½ bottle fruit pectin
3 cups sugar

Stir cordial and sugar together until blended. Put into glass or enameled saucepan and use wooden spoon to stir. Bring to boil over medium heat. Remove from heat immediately and pour in pectin, stirring to blend thoroughly. Pour immediately into hot sterilized jars and seal. Makes about 3 cups jelly.

TUTTI-FRUTTI

A treat that dates from Victorian times and before. Serve over ice cream, or stir into softened ice cream and refreeze to make frozen pudding. Sweet, rich, and very, very elegant.

1 quart brandy   1 quart sugar for each quart fruit
1 quart fruit as it comes into season

Strawberries, raspberries, blackberries, dewberries, elderberries, as well as cherries, apricots, peaches, plums, quince, and whatever else strikes your fancy are welcome to join this mixture.

Clean or pit fruit as necessary. Combine with equal amount of sugar and brandy in a big stone or pottery crock with lid. Stir daily until all fruit has been added, then cover tightly or seal in jars. Store 3 months before using.

RASPBERRY-BLACKBERRY JAM

Not as strong as blackberry jam, nor as sweet as raspberry, this jam strikes a perfect balance in taste and color.

2 quarts red raspberries	9 cups sugar
1 quart blackberries	1 cup water

Mash fruit in layers in deep pot and strain to remove some seeds. Combine sugar and water and boil 5 minutes. Pour into fruit and boil, stirring constantly, until juice sheets from spoon. Pour into hot sterilized jars and seal. Makes about 4 quarts.

STRAWBERRY SUNSHINE PRESERVES

Not entirely foolproof, but the most delicious jam imaginable.

2 quarts strawberries, perfect and fully ripe	2 quarts sugar

Layer strawberries and sugar in deep pot. Let stand 30 minutes, then bring to boil and cook 20 minutes. Stir very gently, just to be sure mixture does not stick to bottom. Spread on platters or trays, cover with glass, and let sit in the sunshine for 3 to 4 days, or until syrup is thick. Stir gently several times during each day and bring inside after sunset. Pack in sterile jars and seal. Makes about 2½ quarts preserves.

STRAWBERRY-RHUBARB PRESERVES

These two come into season together and make a terrific pair.

1 quart rhubarb, cut into small pieces	2 quarts strawberries
8 cups sugar	

Combine chopped rhubarb and sugar, cover, and let stand 12 hours. Then bring quickly to boil and add hulled strawberries. Simmer, stirring occasionally, until thick, about 15 minutes. Pour into hot sterile jars and seal. Makes about twelve 8-ounce jars.

BLUEBERRY JAM

Make this with blueberries that are not quite ripe for extra tang and tastiness.

4 cups blueberries	3 cups heated sugar
½ cup water	

Pick over berries. Put one layer in bottom of deep pot and crush. Put remaining berries on top and add water. Bring to boil and simmer, stirring, over moderate heat until berries are tender. Add sugar and continue to cook, stirring, until a little juice dropped on a cool plate forms a mound and not a puddle. Pour jam into sterilized hot jars and seal. Makes 5-6 cups.

STRAWBERRY MINT JAM

The hint of mint turns ordinary strawberry jam into a food fit for kings. Try it alongside roast lamb or on hot buttered toast.

1¾ cups hulled crushed strawberries	4 cups sugar
2 tablespoons chopped fresh mint, or	2 tablespoons lemon juice
1 tablespoon crushed dried mint	½ bottle liquid fruit pectin

Combine berries, mint, sugar, and lemon juice in deep pan and bring to rolling boil. Remove from heat and stir in liquid pectin. Skim surface and pour into hot sterile jars. Seal. Makes 3-4 cups jelly.

BRAMBLE JAM

Use any of these — raspberry, black raspberry, blackberry, loganberry, dewberry, elderberry. This general rule produces splendid jam. Try also combinations of the fruits.

4 cups berries	3 cups sugar

Crush one layer of berries in bottom of deep pot. Add remaining berries and sugar. Cook, stirring constantly, over moderate heat until mixture boils and sheets from spoon. Pour into hot sterile jars and seal. Makes 5-6 cups.

BLUEBERRY-APPLE JELLY

The apple juice adds tartness and ensures that this will jell.

1 quart blueberries	4 cups sugar
½ cup water	2 tablespoons lemon juice
2 cups apple juice	

Stew blueberries and water for 10 minutes, until fruit is very soft and juicy. Strain through cheesecloth or jelly bag. Measure 2 cups liquid and combine with apple juice, sugar, and lemon juice in large pan. Bring to boil and cook, stirring, until jelly sheets from spoon. Remove from heat, skim foam, and pour into hot sterile jars. Seal. Makes 5-6 cups jelly.

RELISHES

BLUEBERRY-ORANGE RELISH

Pretty in color and delicious in flavor.

2 cups blueberries   ¾ cup sugar
1 orange, seeded and chopped

Mash blueberries. Combine with orange and sugar and heat slowly to boiling. Turn off heat and let stand, covered, 30 minutes. Serve with turkey or chicken. Serves 6-8.

SPICED RASPBERRIES

A spicy spread for cream cheese and crackers, or for buttered toast.

2½ pounds raspberries   1 teaspoon cinnamon
2 pounds sugar, white or light brown ½ teaspoon ground cloves
1 cup cider vinegar   1 teaspoon nutmeg

Combine ingredients in deep pot and bring slowly to boil. Simmer, stirring occasionally, until mixture is as thick as marmalade. Pour into hot sterile jars and seal. Makes about four 8-ounce jars.

CURRIED FRUIT

The curried sauce draws the fruit flavors together. Serve it with hot roast beef, ham, pork, lamb, or poultry.

½ cup butter   2 cups pineapple chunks, fresh or
1 cup brown sugar, packed   canned
1 tablespoon curry powder   2 cups black cherries, pitted, fresh or
½ cup sweet white wine or sherry   canned
2 cups red or black raspberries   1 tablespoon cornstarch mixed with
2 cups pears, fresh or canned, sliced   2 tablespoons pineapple juice

Preheat oven to 350° F. Melt butter in skillet over moderate heat. Stir in sugar and curry powder and cook, stirring, until sugar dissolves. Remove from heat and stir in wine or sherry. Put fruit in deep baking dish, pour sauce over it, and stir in cornstarch mixture. Place in oven and bake 45 minutes. Stir once during baking. Serve piping hot. Makes about 2 quarts.

SPICED BLUEBERRIES

Blueberries never tasted so delectable. These can be used in pie to create a spicy mince-like dessert.

1 cup white vinegar
1½ pounds brown sugar
1 tablespoon whole allspice

1 cinnamon stick
½ teaspoon whole cloves
2 quarts blueberries

Combine vinegar, sugar, and spices and boil 5 minutes. Pour over blueberries in deep pot. Bring to boil and simmer about 5 minutes, until berries are tender and juicy. Cover and let stand 24 hours. Pack berries in hot sterilized jars and pour in syrup. Seal and process 10 minutes in boiling water. Let cool in water bath, then remove and store. Makes about 5-6 pints.

PICKLED BLACKBERRIES

These taste surprisingly good spooned over spice cake, and delicious served with lamb or poultry.

2 quarts boiling water
3 quarts blackberries
1 cup water
6 cups brown sugar

1 cup cider vinegar
2 teaspoons whole pickling spices
2 teaspoons whole cloves
2 teaspoons stick cinnamon, broken

In deep pot, pour boiling water over berries and let stand 5 minutes. Combine 1 cup water, brown sugar, and vinegar and stir until sugar is dissolved. Tie spices in double thickness of cheesecloth. Pour syrup over berries and drop in spices. Bring mixture to boil and boil 10 minutes. Let stand 24 hours, bring to boil, and boil 10 more minutes. Pack in hot sterile jars. Makes six 8-ounce jars.

ELDERBERRY RELISH

Deep purple color, and rich spicy taste. This goes beautifully with game or roast lamb.

5 cups elderberries
1½ cups seeded raisins
1 onion, sliced
1 cup brown sugar
3 tablespoons dry mustard

3 tablespoons ginger
3 tablespoons salt
¼ teaspoon cayenne pepper
1 teaspoon turmeric
1 quart cider vinegar

Combine elderberries, raisins, and onion and chop. Stir in remaining ingredients and bring slowly to boil. Simmer 45 minutes. Pour into hot sterilized jars, seal, and process 10 minutes in boiling water. Cool in water bath, remove, and store. Makes about four 8-ounce jars.

BLACK RASPBERRY CATSUP

A distinctive way to brighten up hamburgers, pork, veal, or poultry.

1 pound black raspberries
½ cup cider vinegar
⅔ cup water
1 cup brown sugar
½ teaspoon cloves

½ teaspoon ginger
½ teaspoon cayenne
1 teaspoon cinnamon
½ teaspoon salt
2 tablespoons butter or margarine

Combine berries, vinegar, and water in saucepan and boil about 5 minutes. Put through food mill or sieve. Stir in remaining ingredients and boil 3 minutes. Cool and serve. Makes 1 pint.

VINEGARS

RED RASPBERRY VINEGAR

Use as the tangy base for a fruit salad dressing or pickled fruits or relishes. And a little added to raspberry jam or jelly while cooking, or poured over raspberry pie before baking, perks up the flavor nicely. (Use the steeped raspberries for jam, relish, or pie. They will have an unusual edge to their flavor which is quite pleasant.)

3 quarts red raspberries, blemish-free

1 quart distilled white vinegar
Sugar

Pick over raspberries carefully and put 1 quart in large glass or crockery bowl. Pour vinegar over them and let stand, covered lightly, 24 hours. Remove berries with slotted spoon and add 1 quart fresh raspberries to the vinegar. Let steep 24 hours. Again remove steeped berries and replace with 1 quart fresh fruit. Let steep 24 hours longer. On the fourth day, remove berries as before. Put liquid into a saucepan and add to it an equal amount of sugar. Bring to a boil and simmer 10 minutes. Pour into hot sterile jars and seal. Makes about 5 cups.

ELDERBERRY VINEGAR

Use in salad dressing for an interesting flavor. Or use as the liquid for pickling other fruits and berries. Best of all, sweeten with honey and heat with whole allspice and cinnamon sticks for a delicious mulled winter warm-up drink.

1 pound elderberries, free of stems
1 pint distilled white vinegar

Sugar

Put cleaned ripe elderberries and vinegar in large glass or crockery bowl. Let stand to mellow for 3 to 5 days. Stir occasionally. After desired time has elapsed, strain fruit into saucepan. Measure juice and add up to 2 cups sugar for each pint of liquid, depending on taste. Bring mixture to boil and dissolve sugar. Continue boiling for 10 minutes. Pour into hot sterile jars and seal. Makes about 2 pints.

BLACKBERRY VINEGAR

Use this vinegar as the base for fruit salad dressings, or as the liquid for making berry relish. (It can also be combined with carbonated water and sweetened to taste to serve as a shrub-like summer beverage.)

1 pound blackberries   Sugar
1 pint distilled white vinegar

Pick over berries and discard any that have bad spots. Place perfect berries in large glass or crockery bowl. Pour in vinegar and cover with a towel. Let steep for 3 to 5 days, stirring occasionally. Strain into a saucepan after desired amount of steeping time. Add up to 2 cups sugar per pint of liquid, depending on preference. Boil mixture to dissolve sugar, and continue boiling for 10 minutes. Pour into hot sterile jars and seal. Makes about 2 pints.

BLACKBERRY WINE VINEGAR

Red raspberries processed according to this recipe also make an excellent vinegar. Using wine vinegar instead of white vinegar gives the end product a slightly richer taste.

6 quarts blackberries   Sugar
1 quart wine vinegar

Put blackberries in large glass or crockery bowl. Pour in vinegar and stir slightly to mix. Place in cool place, cover, and let stand 24 hours. Strain the liquid and measure it into a saucepan. Add an equal amount of sugar. Bring to a boil and stir until sugar is dissolved. Simmer 10 minutes longer. Pour into hot sterile jars and seal. Makes about 2 quarts.

GROWING AND COOKING BERRIES

APPENDIX

Appendix I

STATE AGRICULTURAL EXTENSION OFFICES

From state and county agricultural offices, particularly those in states where large amounts of fruit are grown, the home gardener can obtain soil-testing and nematode-testing services, bulletins and information pamphlets, recommendations for fertilizers and sprays, and general assistance in many departments. Generally all services and pamphlets are free.

States whose offices are apt to be most helpful are California, Illinois, Massachusetts, Michigan, New Jersey, New York, Ohio, Oregon, Pennsylvania, Virginia, and Washington — all states where agriculture is a major industry. But all will know the idiosyncrasies of their own local conditions and how most easily to cope with them on the home-gardening scale.

The national office also offers advice and information. Through it can be obtained bulletins such as:

Commercial Blueberry Growing, Farmers' Bulletin No. 2254, 1973.
Controlling Diseases of Raspberries and Blackberries, Farmers' Bulletin No. 2208, 1975.
Growing Raspberries, Farmers' Bulletin No. 2165, 1979.
Growing Blackberries, Farmers' Bulletin No. 2160, 1979.
Growing Strawberries in the Southeastern and Gulf Coast States, Farmers' Bulletin No. 2246, 1977.
Strawberry Varieties in the United States, Farmers' Bulletin No. 1043, 1979.
Strawberry Culture: Eastern United States, Farmers' Bulletin No. 1028, 1978.
    This pamphlet also comes in editions geared to other regions.
Thornless Blackberries for the Home Garden, Home and Garden Bulletin No. 207, 1975.

All can be ordered from:

Superintendent of Documents
U.S. Government Printing Office
Washington, DC 20402

Address of the national-level agricultural department is:

Publications Division
Office of Communication
U.S. Department of Agriculture
Washington, DC 20250

Order specific bulletins from the former, request additional information from the latter.

States noted as particular strongholds of berry-production also publish excellent pamphlets on care and culture of various berries in their own regions; do not hesitate to write for information.

In many cases, national and state offices will levy a slight fee for pamphlets; they will be sure to let you know if this is the case.

Appendix II

SOURCES OF PLANTS AND CATALOGS

The list offered here is not intended as an endorsement or recommendation, but rather as a suggestion and a convenience. What is recommended is that you acquire catalogs from a variety of sources, examine them and their offerings carefully, and make your own choices on the basis of local conditions and common sense.

Ahrens Nursery R.R. 1 Huntingburg, IN 47542	All types available
Alexanders Nurseries P.O. Box 309-Y Middleboro, MA 02346	Blueberries
A.G. Ammon Nursery Box 488 Narnegat Road Route 532 Chatsworth, NJ 08019	Blueberries
Atlantic Blueberry Co. Galletta Bros. & Sons Blueberry Farms Hammonton, NJ 08037	Blueberries
Boston Mountain Nurseries Rte. 2 — Hwy. 71 Winslow, AR 72959	Bramble fruits, 2 strawberry varieties
Bountiful Ridge Nurseries Princess Anne, MD 21853	All types
Brittingham Plant Farms 2538 Ocean City Road Salisbury, MD 21801	Blueberries, raspberries, and especially strawberries
Buntings' Nurseries, Inc. Selbyville, DE 19975	Blueberries, brambles, and especially strawberries

Burpee Seed Co. Warminster, PA 18991	All types
Carroll Gardens P.O. Box 310 444 East Main St. Westminster, MD 21157	Limited selection of blueberries
The Connor Co. P.O. Box 534 August, AR 72006	Strawberries
C & O Nursery 1700 N. Wenatchee Ave. P.O. Box 116 Wenatchee, WA 98801	Blueberries, brambles
Emlong Nurseries, Inc. Stevensville, MI 49127	Some blueberries, elderberries, brambles
Farmer Seed and Nursery Co. 818 N.W. 4th St. Faribault, MN 50021	All types
J.M. Faubus Faubus Berry Nursery Star Route 4 Elkins, AR 72727	Brambles, strawberries
Henry Field Seed & Nursery Co. 407 Sycamore St. Shenandoah, IO 51602	Most types
Finch Blueberry Nursery Route 1, Box 341 Bailey, NC 27807	Rabbiteye blueberries
Dean Foster Nurseries Hartford, MI 49057	Most types
Gardens of the Blue Ridge P.O. Box 10 Pineola, NC 28662	Wild strawberries
Gurney Seed & Nursery Co. Yankton, SD 57079	All types
Hartmann's Plantation Inc. Patrick J. Hartmann & Sons Route 5 So. Haven, MI 49090	Extensive blueberry selection
Hastings Box 4274 Atlanta, GA 30302	All types — southern varieties
Interstate Nurseries Hamburg, IO 51644	Most types

Merrill W. Jewett Hyde Park, VT 05655	Limited selection strawberries, everbearing raspberries
Rice Nurseries Geneva, NY 14456	All types
Riverview Nursery Co. Highway 55, Rt. 3 McMinnville, TN 37110	Raspberries
Savage Farms Nurseries P.O. Box 125, SF 1234 McMinnville, TN 37110	Most types
R.H. Shumway, Seedsman Rockford, IL 61101	Some types
Smith Berry Gardens Ooltewah, TN 37363	Strawberries
Spring Hill Nurseries 110 West Elm St. Tipp City, OH 45366	Most types
Stark Bros. Nurseries Louisiana, MO 63353	Strawberries, blueberries, brambles
Stern's Nurseries, Inc. Geneva, NY 14456	Elderberries
Waynesboro Nurseries, Inc. Waynesboro, VA 22980	Strawberries, blueberries, brambles
White Flower Farm Litchfield, CT 06759	Fraises des Bois, French & Russian varieties
Wyatt-Quarles Seed Co. Box 2131 Raleigh, NC 27602	Strawberries

INDEX

ABOUT THE AUTHOR

If *Mary Cornog* lived in England rather than *New* England, she would be referred to in print as: Mary Wood Cornog, A.B., M.A., Ph.D. "A.B." because that's what the English call our American B.A., which Mary got from Wellesley College; she was awarded her Master of Arts in Greek Language and Literature by Columbia University, and the PhD. by Boston University in Classical Studies — Greek again, which she speaks, reads, writes and teaches fluently. She has also taught Greek, Latin, Greek literature in translation and Ancient and Medieval History at The Dublin School, in Dublin, New Hampshire, where her husband is Headmaster.

Why is a classical scholar writing about gardening and cooking? Well, Mary was brought up and still lives in the country and has been gardening and cooking all her life — starting long before she discovered classical Greece at college. She grows all her own vegetables and flowers, and, of course, her own berries! In addition, she writes for *Yankee* Magazine, *The Old Farmer's Almanac,* and other publications, cooks for her family and visiting notables, and takes care of a menagerie of one dog, one horse, five cats, two rabbits, and five chickens. Daughter Sarah, aged eight, is her favorite berry-picking companion.